Brooklyn's Bushwick - Urban Renewal in New York, USA

Raymond Charles Rauscher • Salim Momtaz

Brooklyn's Bushwick - Urban Renewal in New York, USA

Community, Planning and Sustainable Environments

Springer

Raymond Charles Rauscher
University of Newcastle
East Gosford, NSW, Australia

Salim Momtaz
University of Newcastle
Ourimbah, NSW, Australia

ISBN 978-3-319-38121-3 ISBN 978-3-319-05762-0 (eBook)
DOI 10.1007/978-3-319-05762-0
Springer Cham Heidelberg New York Dordrecht London

Cover image: Renewal in Brooklyn Neighborhoods: Bushwick row houses (Cedar Street, 1982)

Photograph courtesy of Ray Charles Rauscher

Printed on acid-free paper

Springer is part of Springer Science+Business Media (www.springer.com)

This book is dedicated to six universalist writers spanning nearly 100 years (birthdays between 1819 and 1915). These individuals were either born in, or lived part of their lives in, New York City, including: Walt Whitman, Henry Thoreau, Helena Blavatsky, Henry George, Henry Miller, and Arthur Miller (Plate 1). Universalist thinking promoted liberating philosophies and environmental perspectives within their writings.

Walt Whitman (1819–1892) was born on Long Island (now the Walt Whitman Center, West Hills, Long Island). He was educated in Brooklyn and lived there for a number of years. He was editor of the Brooklyn Eagle in 1846 and published his famous Leaves of Grass in 1855. He also wrote Drum-Taps (1865), Democratic Vistas (1871), Passage to India (1872) and Specimen Days (1882). Whitman also lived, until his death, in Camden, New Jersey (1873–1892).

Whitman, were he alive, would have covered the September 11, 2001, terror attack on the World Trade Centre in Lower Manhattan (Plate 2). Whitman at one stage

Plate 1 Universalist Writers Associated with New York City. *Left* to *Right*: Sources: Mark Van Doren, Ed.; Theosophical Society; Henry George Society, NYC; Encyclopedia Britannic; Wikipedia

Whitman walked Lower Manhattan observing the American resilience after the terrorist attack 9/11
Whitman's heart always burned for the City of mixed national groups forging new destinies
He walks today as he reminded us in his poems he would be there a thousand years later
Today he sees his New York flickering with candles and grasping to regain its lost heart

Whitman tapped the spirit of people in their intimate life and work engagements
He saw the fireman, office worker, street vendor and boot maker as the City itself
Whitman spoke of the brisling muscle and misty eyed romantic hearts filling New York
New York for Whitman was the crucible of all America fusing into one

Whitman inspired all Americans to see their deeds as a national tune in stepping forward
Irving Berlin reflected this gift in his embodiment of national pride in God Bless America
Arthur Miller translated humankind's constant search for sanity and compassion
As Easter follows Good Friday, America will see the Whitman spirit fully regained

Plate 2 Walt Whitman Looks On (Source: Ray Rauscher, Sydney, 2013 (unpublished))

of his journalism life worked in Lower Manhattan not far from what is now seeing new buildings rise at Ground Zero. The New Yorker's and American's resolve to heal and understand the impact of such a tragedy would have impressed Whitman. Out of the rubble with nearly 4,000 lives lost, the idea of 'fortress' America as impenetrable was also lost. Whitman was always the observer of the American spirit and its democratic ways. He championed equality and justice as everyone's birth right, a reminder of the gift Whitman gave to America. A poem to Whitman is a thank you for his contributions.

Whitman would have admired Henry Thoreau's writings. Thoreau studied and wrote about nature and its relation to the human condition, especially in his famous book Walden (1846). He lived for several years in Staten Island, New York City, where he stayed with fellow writer Ralph Waldo Emerson (1803–1882) who wrote Nature (1836). Thoreau associated with the idealist philosophies advocated by Emerson, Fuller, and Alcott. They held high the beliefs of an ideal state of life going beyond the physical and empirical, with personal intuition as vital to life's decisions.

While Whitman was working in New York City (middle-to-end 1800s), Madame Helena Blavatsky (1831–1891) (Russian domiciled in New York City for a time) launched the progressive thinking Theosophical Society in the City (1875). The Society's three principles (popular with many New Yorkers at the time), being to: form a nucleus of the universal brotherhood of humanity without distinction of race, creed, sex, caste, or color; encourage the study of comparative religion, philosophy, and science; and, investigate the unexplained laws of nature and the powers latent in humanity.

In New York City at this time (middle-to-end 1800s) there was Henry George (1839–1897) (spent early years in Pennsylvania), a seer and philosopher of urban land use reform. George wrote about people's right to a better share in the prosperity that urban areas were creating. He saw degradation forming as affluence grew, and thus wanted to find out why this happens. Whitman would have been aware of George's work Our Land and Land Policy (1871). Here George examines land speculation in cities, showing there was too little return to governments (the people) in land transactions. He later wrote his popular work Progress and Poverty (1879), considered one of the greatest works of the English language (translated into almost every language). The perplexing question George struggled with in his life (poverty co-existing with abundant wealth) was ever evident in New York City. The forthrightness of George (bearded and sharp of eye) would have presented a formidable figure challenging those in city government to address poverty and land use reforms . He ran for mayor of New York City, polled second and ahead of Theodore Roosevelt (mayoral race was won by Abram Stevens Hewitt). The Robert Schalkenbach Foundation based in New York City (www.schalkenbach.org) (13 May 2013) claims "the men who believed in what George advocated called themselves disciples." George is buried at Green-Wood cemetery, Brooklyn.

The fourth person, who came later, within the universalist writers was Henry Miller (born 1891, New York City). Miller was brought up in Brooklyn and wrote about his childhood experiences in Black Spring (Miller 1936). He was always ready to speak in candor about the urban condition and injustices midst the chaos in places such as New York City. Miller had a gift of comedy in writing about people facing the challenges of city living, especially the cost in human terms of mechanization and commercialization. Henry Miller wanted to get to the roots of the American nature and experience, so he travelled across the country to gain the knowledge of American life. The Air-Conditioned Nightmare (Miller 1945) proved an excellent discourse on American life at that time. This work was reflective of the other greats of liberal thinking and writing noted above, Whitman, Thoreau, Blavatsky, and George. History proved there would be many other writers on these subjects along writers' footsteps on the streets of New York City.

The fifth and final person who would have had a good deal in common with all those above, though from a different era, is Arthur Miller. He was born in New York City in 1915. Miller had a social awareness growing out of his experiences of the Depression years (1930s). He could thus portray the insecurities of human existence within urban settings, and did so in Death of a Salesman (Miller 1949). Here was a book that focused on the urban man, facing the challenges of a struggling family in hard times (reflective of many Brooklyn families in the 1940s–1950s).

Book Overview

The book offers a case example model of the urban communities of Bushwick and Northeast Brooklyn, New York City. The story of Bushwick is the central starting point, given its recovery from decades of decline. The book commences (Chap. 1) by examining the urban history of Bushwick and Northeast Brooklyn up to the 1960s. Topics include: famous people in Brooklyn's past; disappearance of native American Indians; significance of the Battle of Long Island (1776); Old Brooklyn's Towns pre-1854; the emerging of the City of Brooklyn (1850s); and, finally Bushwick's changing street patterns, and rise of industry, institutions, commercial areas, services and churches.

Bushwick planning, from 1970s to current times, is reviewed in Chap. 2. This period includes the demise of Bushwick that culminated in the catastrophic Bushwick arson fires and looting in the New York City blackout of 1977. In this incident, a large section of Bushwick housing was lost as well as scores of businesses (extent of loss examined in Chap. 2). The reader is introduced to the recovery of Bushwick, including results of research coming from author's field trips to Bushwick and Northeast Brooklyn over several years. The recovery of Bushwick, particularly, in the 1980s, 1990s and early 2000s is reviewed. Finally, examples of 'green design' within developments and newly emerging principles of 'sustainable urban planning' (SUP) are commented on.

Chapter 3 steps back from the immediacy of Bushwick and Northeast Brooklyn to examine urban change from a theoretical perspective. The chapter first examines the movement of urban planning from early practices to current applications of sustainable urban planning (SUP). The advances in sustainable urban planning (SUP), and how it could be applied to Bushwick and Northeast Brooklyn (and wider New York City, or anywhere) is examined. The chapter covers: defining sustainability; reviewing changes to urban planning practice; introduction to ecologically sustainable development (ESD); and, the adoption of sustainability criteria within urban planning. To understand sustainable urban planning (SUP), the chapter reviews the historical changes in urban planning schools over more than half a century; three schools are examined. Finally, a critique is completed on ESD and its application within urban planning, from international level to local.

With an awareness of sustainable urban planning (SUP), the next chapter (Chap. 4) examines New York City planning instruments and the innovative structure of local community boards. A key urban planning tool used by the City of New York (the administration), the 197-a Plan process, is examined. This planning process is a community planning system established by the City of New York in 1975. The 197-a Plan process was created to assist planning and revitalization of areas like Bushwick and Northeast Brooklyn. In addition, the urban planning roles of local City of New York chartered 'Community Boards' (CBs) are examined, especially around the planning instruments as the 197-a Plan process.

The application of the brownfields development planning and the 197-a Plan process (including engagement of Community Boards) to urban renewal projects in Bushwick and Northeast Brooklyn are examined in Chap. 5. In addition, to gain a comparative look at urban planning beyond these areas, the chapter looks at two prime Brooklyn neighborhoods. These are Brooklyn Downtown and Southeast Brooklyn (incorporating Brooklyn's historical recreation area of Coney Island). Conclusions are drawn on the applications of these urban planning instruments (and role of Community Boards), including whether principles of sustainable urban planning (SUP) (Chap. 3) were applied.

A vital ingredient in successful urban planning (local, regional, city or state) is the education of citizens in how planning works and how citizens can be involved in that planning (Chap. 6). This education encompasses, for example, students understanding neighborhoods and how they function. Education authorities are increasingly aware of the value in exposing students to the theory and practice of urban planning, especially as that planning affects the neighborhoods those students live in. This chapter thus looks at a number of innovative education programs, using the Bushwick neighborhood as a case example of changing education approaches.

A select number of high schools experimenting in teaching urban planning as a subject, including practical student work within neighborhoods, is examined. The chapter examines aspects of this education approach and comments on exposing students to urban planning experiences. The chapter reflects back (in examining the relevance of today's classroom in preparing students for an urban future) on: a. lessons from the history of urban planning in Bushwick and Northeast Brooklyn (Chaps. 1 and 2); b. theory and practice of the emerging sustainable urban planning (SUP) (Chap. 3); and, c. structures and applications of the City of New York planning instruments (i.e. 197-a Plan process) and Community Boards (Chaps. 4 and 5). Conclusions are reached on the value to students to exposure to new education experiments. This includes the value of students' engagement in practical exercises in neighborhoods. Finally, the value of these education experiments to students' academic futures is commented on.

Chapter 7 looks at common themes identified and conclusions reached within the chapters presented. The chapter, from these conclusions, examines future directions of urban planning (with hindsight of Bushwick, Northeast Brooklyn,

Downtown Brooklyn and Southeast Brooklyn planning). Finally, the directions of sustainable urban planning (SUP) and the means of advancing SUP are commented upon. The reader, in pursuing further studies of the subject, is reminded of the availability of web-based reference resources on matters raised within the book and generally on the subjects of sustainable urban planning (SUP) within Appendix 3.

Acknowledgements

We acknowledge the contributions of residents of Bushwick and Northeast Brooklyn, past and present. The resilience of all Bushwick residents and workers contributed to the compiling of this story. To gain a historical perspective of changes in Bushwick through the 1950s and 1960s that may have flowed into the 1970s, We set up a research based group email. The group (seven in all and living in different parts of America) were from the Bushwick High School Class of 1961 (see Chap. 6 and Appendix 1). We also liaised with Bushwick High School (which in 2003 became three schools in one campus, including the Academy of Urban Planning and School of Social Justice) (Chap. 6). *The Bushwick Brooklyn USA Story* (Rauscher 2010, unpublished) (CD copy) was personally presented in 2010 to: Monique Darrisaw (Principal), Adam Schwartz (Lecturer), and Josh Lapidus (Program Manager) from the Academy of Urban Planning; Amanda Eyrich-Daly (Director of Education, Brooklyn Center of Urban Environments) (center ceased in 2009); Leilani Dawson, Education Unit, Brooklyn Historical Society; and, Dr. Guy Hehner, Dir of Education, Henry George School of Social Science, 30th St. Manhattan.

Organizations and their representatives I and Dr. Momtaz wish to thank for assistance (between 1979 and 2010) include (a–z): Artist-Photographer Teacher (Meryl Meisler); Arts in Bushwick; Brooklyn Community Board 1 (Noah Pfefferbilt); Brooklyn Community Board 4 (Nadine Whitted); Bushwick Courier (Aaron Short); Bushwick History (John Dereszewski); Center for Human Environments (City Univ. Of New York) (Sanford Gaster, Roger Hart); City of New York, City Council Member, Brooklyn (Diana Reyna) and community liaison officer Francisco Mercado; City of New York: City Planning Dept (Beverly Tallevera); Dept of Housing, Preservation and Development (Eva Brown, Jack Hamer, Herbert Siegel; Cassandra Vernon, Rubin Wolf); Columbia University, Urban Planning (Lionel McIntyre); East Broadway (Brooklyn) Business Improvement District (Matthew Schneiderman); Greater Ridgewood Historical Society, Onderdonk House; Green Maps, Manhattan; Historic District Council; Hunter College of the City University of New York (CUNY) (Prof. Tom Angotti); Landmarks Preservation Commission (Marianne Malec); Municipal Arts Society (Juan Camilo Osorio); Neighborhood

Preservation Center (Felicia Mayro); New York Landmarks Conservancy (Roger Lang); Pratt Institute (Ron Shiffman); Project for Public Spaces; Ridgewood Bushwick Senior Citizens Council (Wesley Hitner); Robert Schalkenbach Foundation (Ted Gwartney, Mark Sullivan); St Barbara's Catholic Church (Monsignor John Powis); and, Transport Alternatives, Manhattan; Walt Whitman Centre, West Hills, Long Island; Waterfront Alliance (Carter Craft).

Special thanks to others who through emails, blogging and internet postings have contributed. Acknowledgment is made to contributions by Bushwick Class of 1961 graduates: Mari Degener (nee Grossman); Susan Ferrandiz; Betty Lee (nee Wasielewski); Tina Morello; Elaine Portnoy (nee Grossman); and Patricia Ward. Also thanks to Joe Salvia (early Bushwick Highschool and Grover Cleveland Class of 1956) with mapping assistance of early Bushwick. Within Australia, thanks to Habitat Association for Arts and Environment Inc. for putting the earlier Bushwick Story on line (John Cristoforou, Ruth Dickson, David Holland, Megan Hitchens and Margaret O'Toole (both proof reading the book), Diane Rauscher, and, public officer Kevin Armstrong). Thanks also to IT support from Kevin and Anna Rochford, with photo review by Brooklyn Rochford. In addition, secretarial assistance from Univ of Newcastle School of Environment and Life Sciences (SELS), Pam, Nicole and Fiona. Finally, thanks to Maree Wheelahan, for project encouragement.

On trips to Bushwick, Greenpoint and Williamsburg, We were assisted by (and thank) the following Rauscher family members, all of whom lived in Brooklyn their whole lives or part thereof: Grace Rauscher (the late); Henry Rauscher (the late); Arthur E. Rauscher (the late); Gracie Riccio (nee Rauscher) (the late); Arthur Rauscher; Joyce Martin (nee Rauscher); Philip Rauscher; and Joseph Riccio (the late). Finally, thanks to all members of both authors' families, Rauscher and Momtaz. It is the total of all these contributions that makes a book worth the effort.

Governments are always working behind the scenes and results of efforts can take decades to see. The City of New York's Dept of Housing and Preservation played (and continue to play) a major part in Bushwick's and Northeast Brooklyn's recovery. Behind good planning is usually good elected officials and staff. Thus all City of New York's past and present mayors, Brooklyn Councilors, Brooklyn Borough controllers, and staff deserve acknowledgment for efforts to face the tough urban challenges that renewal of Bushwick and Northeast Brooklyn present. Mayor Lindsay, in the 1960s, was one of the Mayors of the City of New York to see the problems looming in Bushwick and Northeast Brooklyn. His successors, including immediate past Mayor Bloomberg and all concerned, continued the work that sees an ever emerging renewal in this part of the City. We wish the new Mayor (109th for the City) Bill de Blasio, new Councilors and the administration the best of luck in their dedicated work ahead.

Contents

Abbreviations

197-a Plan	City of New York community planning charter
BID	Business improvement district
CB	Community board
DPNYC	Department of Planning, New York City
ESD	Ecologically sustainable development
ICLEI	International Council for Local Environmental Initiatives
IUCNNR	International Union for Conservation of Nature and Natural Resources
LA21	Local Agenda 21
LG	Local government
NGO	Non-government organization
NYC	New York City
RBSCC	Ridgewood-Bushwick Senior Citizens Council
SUP	Sustainable urban planning
TBL	Triple bottom line
UN	United Nations
UNESCO	United Nations Educational, Scientific and Cultural Organization
WWF	World Wildlife Fund

List of Appendices

List of Figures

List of Plates

Chapter 1
Bushwick and Northeast Brooklyn: Formative Years to 1960s

1.1 Introduction

This chapter examines the history of urbanization of Bushwick and Northeast Brooklyn. The book starts with comments on how these areas were part of the European 'new world'. It was a place of hope and inhabited by Native American Indians. The chapter notes some highlights of the evolution of Bushwick and Northeast Brooklyn to a rural settlement and to an eventual urbanized one, up to 1960s. A brief look at key events in that urban transition includes comments on (and provision of maps and pictures): pre-European settlement; position of the indigenous community of American Indians; and, the independence movement away from England by Americans (new settlers).

The profound effect of the Battle of Long Island (1776) (also referred to as the Battle of Brooklyn) as a turning point in American history is commented on. The reader is then introduced to the era of the Dutch towns of Brooklyn in the 1850s. Taking Bushwick as a case example of urbanization within Northeast Brooklyn, key features of Bushwick are examined. These features include: commercial facilities, industry, public institutions, and churches. Finally, moving beyond the formative years, the chapter examines urban life in Bushwick in the 1940s–1960s. The next chapter looks at Bushwick planning from the 1970s to current times.

1.2 Pre-European Settlement

Any urban history in America starts with the Native American Indians. A journey by Adrian Block (1613–1614), a Dutchman, resulted in a 1635 map of the Long Island (New York) area (Plate 1.1) (via The Richmond Hill Historical Society) adapted from *Our Long Island* (Manello, unknown publishing date). The map shows the various Native American Indian tribes known to have been in the area around 1643. These included (a–z): Canarsie, Corchaug, Manhasset, Massapequa, Matinecock,

R.C. Rauscher and S. Momtaz, *Brooklyn's Bushwick - Urban Renewal in New York, USA: Community, Planning and Sustainable Environments*, DOI 10.1007/978-3-319-05762-0_1,

Plate 1.1 Long Island Native American Tribes 1635 (Source: The Richmond Hill Historical Society, NYC 2013)

Merrick, Montauk, Nissequoge, Patchoag, and Rockaway, Seatauket, Secatoag, and Shinnecock. Those familiar with Long Island would recognize the many districts of Long Island that bear the names (or similar) to these Native American Tribes. The Shinnecock maintain a significant reservation on the eastern (southern end) of Long Island (shop open to the public).

On a negative historical note, the Historical Society notes: "Trouble with the Dutch was one of the factors which eventually drove the Indians from Long Island. Another was smallpox epidemic in 1658, which reportedly killed two thirds of the tribes in the area. Moreover, the influx of white settlers and the resulting expansion of farmland drove animals away, and the Indians who were hunters migrated to the mainland in pursuit of game. By 1741, it was estimated that only 400 natives remained on the island. By the time of the American Revolution in 1775, Indians were a rare sight on the island, having been driven away from their beloved 'Paumanok" (Indian name for Long Island, land of tribute'). The Historical Society reached to the Henry Wadsworth Longfellow *Song of Hiawatha* (1855):

I have seen it in a vision
Seen the great canoe with pinions,
Seen the people with white faces,
Seen the coming of the bearded
People of the wooden vessel
From the region of the morning.
From the shining land of Wabun
'Gitchi Manto', the Mighty
The great spirit, the Creator
Sends them hither on his errand,
Sends them to us with a message

1.3 Revolution Time and Brooklyn

In June 1776 British troops gathered on Staten Island anticipating trouble with the Continental Army of Americans. Commander in Chief George Washington brought the army (about 11,000 men) to New York City, expecting a British invasion. By July that year the Declaration of Independence was presented to Congress by Thomas Jefferson and signed on the 4th July. By August of that year 20,000 British troops sailed across the narrows and landed in Gravesend Bay to the south east of Flatbush. A map (Plate 1.2) depicts roughly the travel lines of English troop movements and forts manned by the Americans (as the settlers and revolutionaries were known).

The British passed within days into Bedford Village, south of Bushwick. The Americans were forced to fall back to a line of earthen forts around Brooklyn Heights. Other forts in Brooklyn at the time were: Defiance; Stirling; Putnam; Corkscrew; Lawrence; and Box. Within days the bulk of the American forces retreated across the East River (left in Plate 1.3). The whole battle on 27 Aug 1776 became known as the Battle of Long Island (or Battle of Brooklyn). Over 2,500 American men were killed, wounded or captured during this battle.

The Village of Williamsburg was incorporated within the Town of Bushwick in 1827, but later became the City of Williamsburg. Note that there was both the Town of Bushwick and the hinterland of Bushwick, a large part of that became part of the borough of Queens. Bushwick had its own Town Mayor, town hall, taxes and works programs. To a degree the establishment of the current Community Boards, to be

Plate 1.2 Battle of Long Island 1776 (Source: Dept of the Interior (Stiles) 2013)

Plate 1.3 Old Brooklyn's Towns Pre-1854 (Source: City of NY Archives 2013a)

outlined in Chap. 4, is a return to 'the local town hall' that towns like Bushwick had. By 1854, both Bushwick and Williamsburg were annexed by the City of Brooklyn and became collectively the 'Eastern District of Brooklyn'.

1.4 Dutch Towns

The name 'Boswijck' (Bushwick) means 'little town in the woods', reflecting the wooded areas at the time. Bushwick was one of six Dutch towns as founded in 1638 and chartered by Peter Stuyvesant in 1661 (Plate 1.3). These towns, in addition to Bushwick, included (clockwise): Brooklyn, Flatbush, Flatlands, Gravesend, and New Utrecht.

1.5 Bushwick in 1850s

Bushwick has always been a working class neighborhood, originally popular with immigrant groups. For over a century (early 1700s to 1800s) the neighborhood consisted of farms. The largest immigrant group arriving in the 1840s was from Germany and worked mostly in industry. By the twentieth century, Bushwick was a center for shipping and manufacturing. Soon immigrants in large numbers arrived from Italy and other European countries. A further wave of Hispanic immigrants was to arrive in the 1960s and 1970s. Bushwick thus represents a cross section of the American immigrant history of nation building. That trend has over the years extended to an even wider group of national backgrounds of immigrants settling in New York City.

The growth of New York City and the positioning of Bushwick and other towns in the early 1850s is illustrated in Plate 1.3. A prominent band of green escarpment ran from east to west separating the urban from the rural areas, containing Flatbush and New Lots for example. Within Brooklyn major streets such as Fulton St., Atlantic Ave, and Broadway were already thoroughfares by the 1850s. Important localities at the time (Plate 1.4) included: (1) Greenpoint and Williamsburg; (2) Flatbush; (3) New Lots; (4) Downtown Brooklyn; (5) Gowanus Bay; and, (6) Bushwick.

The development of Bushwick, influenced by the Dutch and English in particular, resulted in a grid pattern of streets (Plate 1.5). This pattern emerged from the years of farm paddocks (as noted above) being gradually urbanized.

1.6 Bushwick in 1900s

By the early 1900s the area of urban Bushwick had filled all the paddocks (Plate 1.6). The neighboring areas also became urbanized, including (1–3 in Brooklyn and 4–6 in Queens): (1) Bedford Stuyvesant; (2) Williamsburg; (3) Greenpoint; (4) Ridgewood; (5) Glendale; and, (6) Middle Village in Queens.

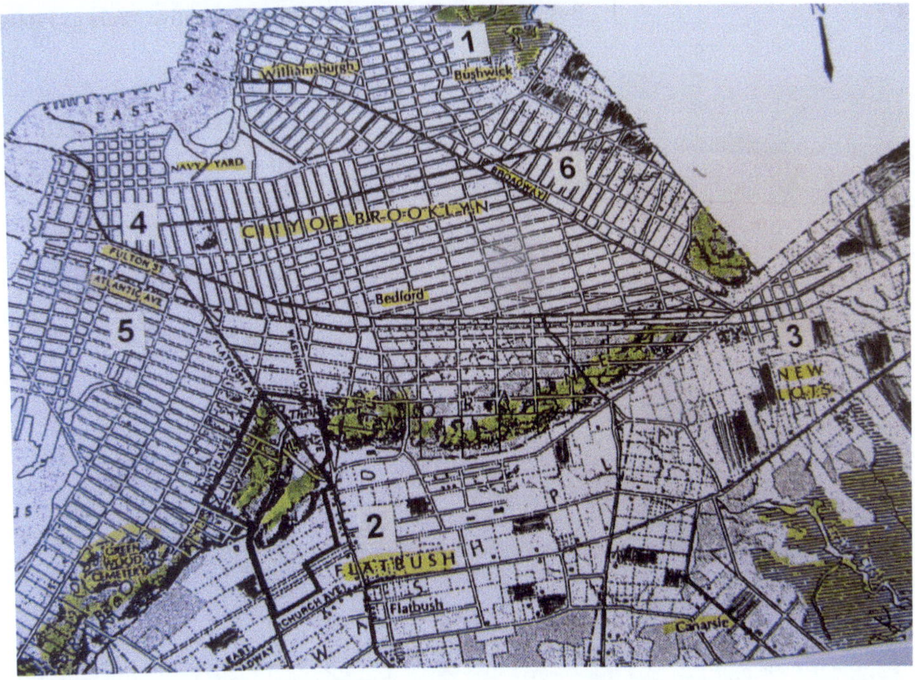

Plate 1.4 City of Brooklyn 1850s (Source: City of NY Archives 2013b)

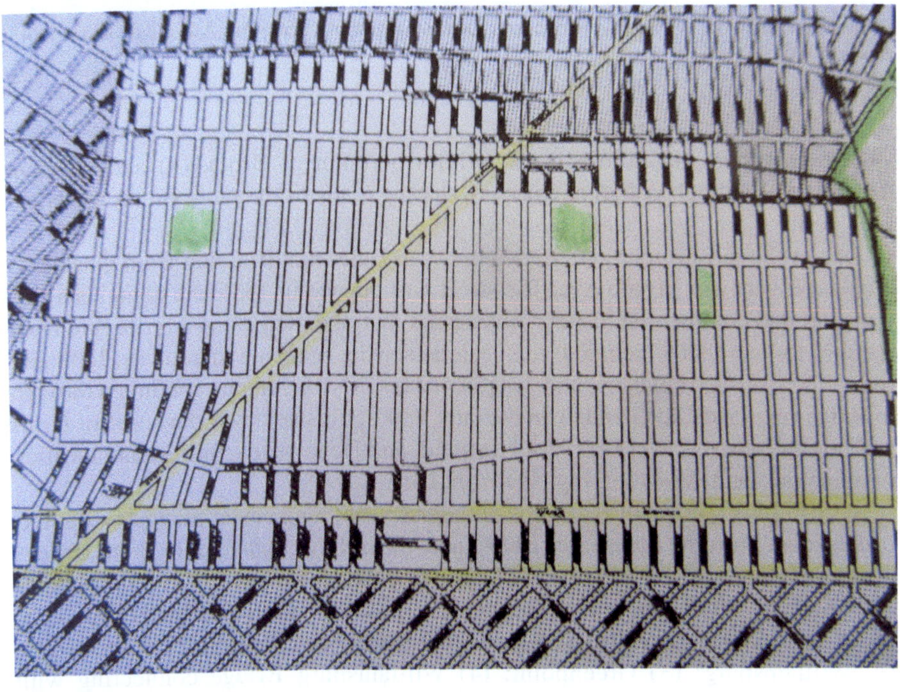

Plate 1.5 Early Grid Streets of Bushwick (Source: City of NY Archives 2013a, b)

Plate 1.6 Bushwick within Brooklyn and Queens Boroughs (Source: American Automobile Assoc 2000)

The neighborhood of Bushwick (within borough of Brooklyn) and adjacent Ridgewood (within borough of Queens), sharing many of the same streets, is shown in Plate 1.7.

One of the most important arteries in urban Bushwick (running from Queens to the East River) was Myrtle Ave. The street was opened in 1835 and contains a rich transport and commercial history. Businesses that operated along Myrtle Ave from the mid-to-late 1800s included: tea dealers; sewing machine merchants; lightning rod sellers; sausage butchers; and, wagon makers. Myrtle Avenue continued through to Ridgewood and Old Glendale in the later 1800s. This area was the home of farms and homesteads (i.e. Wyckoff, Van Nostrand, and Debevoise families) and location of numerous cemeteries serving New York City.

Here in Plate 1.8 are three photos of the busy shopping center along Myrtle Ave, Ridgewood. The time is the early 1900s (note the trolley car in middle photo, always popular in connecting Bushwick life with that of Ridgewood).

A look at Brooklyn, with a perspective on Manhattan and further afield is provided in Plate 1.9. The map depicts in this order key localities: (1) Bushwick; (2) Williamsburg; (3) Greenpoint; (4) Williamsburg Bridge connecting with Manhattan; (5) Manhattan Bridge; and, (6) Brooklyn Bridge.

Plate 1.7 Bushwick neighborhood (Source: American Automobile Assoc 2000)

Plate 1.8 Early 1900s Myrtle Ave Shopping – Ridgewood (adjacent to Bushwick). *Left* to *Right*: Shopping Street; Trolley Bus; Cars and Bicycles (Source: New York City Archives 2013a, b)

A trip through Bushwick, Williamsburg and Greenpoint streets (many connecting to Bedford-Stuyvesant, Ridgewood, and Glendale) yields many famous American historical figures. Some prominent streets include: Dekalb (revolutionary war officer); Knickerbocker (early American writers club active in NYC); Cypress (prolific tree in this area); Cedar (common timber evergreen conifer); Myrtle (species of evergreen shrubs); Stuyvesant (Dutch Colonial governor who tried to resist the seizure of NYC by the British colonials and declared NYC as New Amsterdam in 1653); Jefferson (Thomas Jefferson, 3rd President of USA and principal author of the

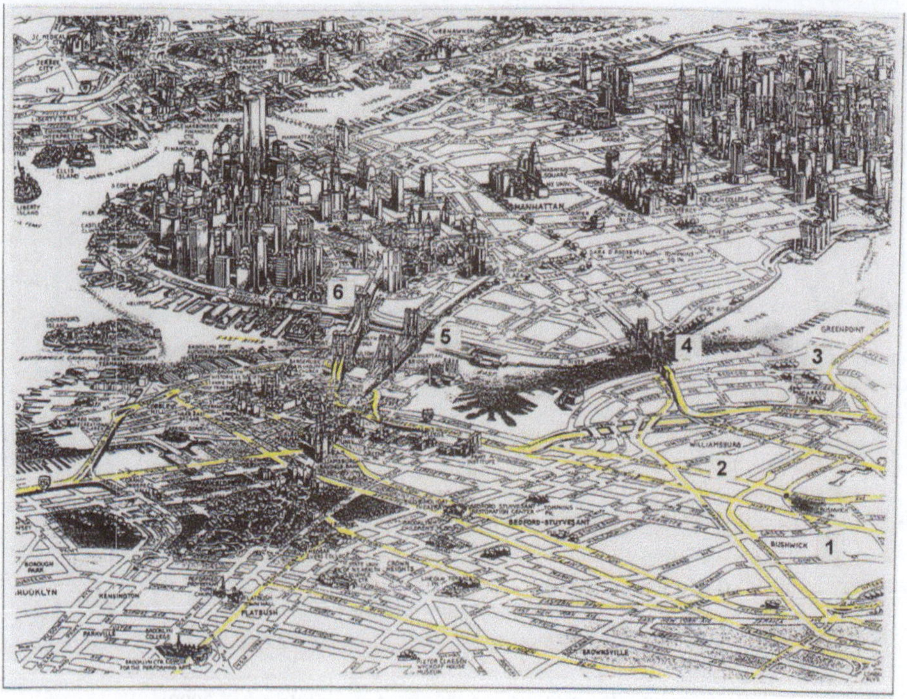

Plate 1.9 North Brooklyn and Manhattan 2000 (Source: unknown)

Declaration of Independence); Sumner (Charles Sumner, statesman of the Civil War who championed human equality and crusades from prison reform to world peace); Franklin (Benjamin Franklin, author, inventor, scientist and statesman); Emerson (Ralph Waldo Emerson, poet and lecturer on transcendental thinking); and, Washington (George Washington, first President).

A further look at these streets of Bushwick, Williamsburg and Greenpoint reveals other prominent names such as: Clinton (George Clinton, soldier, Governor of NY and Vice President of USA); Vanderbilt (Cornelius Vanderbilt, 'the commodore' founded steamships and railway lines); Fulton St (Robert Fulton, inventor brought steam boating to commercial success); Irving (Washington Irving, first American man of letters and father of American literature); and, Gates (Horatio Gates, led a victory over the British in 1777 that turned the tide to the American revolutionaries).

Other street names of special historical importance include: Penn (William Penn, Quaker leader and advocate of religious freedom); Roebling (John Roebling, designer of the Brooklyn Bridge); Kent (James Kent, jurist whose decisions shaped common law in the formative years of US); Evergreen (original Bushwick Ave promenade); Flushing (nonconformist who lived in Queens and developed a Quaker Centre); Kosciusko (Tadeus Kosciusko was an officer and statesman in the US War of Independence); Lafayette (fought with the American colonists and was influential in

Plate 1.10 Under the Broadway Elevated (Source: *History of Williamsburg*. Brooklyn Historical Society 2000)

Under the 'el' would give you the experience of a life
The thunder and rattle of trains grabbed your heart and breath
The dark forborne spaces below were of shadows and light
The speckled pavement was like artwork as in fancy flight

Ascend those cast iron cold steps in the chilled winter bite
The transit lady bundled in woolens and beanie hat slips a token your way
You push the stubborn turnstile and get jolted onto the windy platform
Wavering in reflecting sun, the train screeches to stop, thanks New York Transit

Plate 1.11 The Bushwick El (Source: Ray C. Rauscher, Sydney 2013 (unpublished))

the French Revolution); Van Buren (Martin Van Buren, 8th President of USA and a founder of the Democratic Party); Madison (James Madison, 4th President of USA and founding father of he US Constitution); Putnam (Gen Israel Putnam, American revolutionary war hero); and, Hancock (John Hancock, American revolutionary leader and signature of the Declaration of Independence).

One picture that tells an aspect of Brooklyn's early public transport is the NYC elevated train lines ('the 'el'). This is illustrated here in a 1920s picture (Plate 1.10) and in a poem encompassing an encounter with 'the Bushwick 'el' in the 1950s (Plate 1.11).

There are many references that provide a glance at urban Brooklyn of yesteryear.

To visit one's old neighborhood can be a daunting experience as familiar areas and buildings you once knew may be gone. Memories of streets, with intimate understandings are etched deep within most people. Many people thus maintain a drive to re-connect the places and people of the past.

Bushwick, Williamsburg and Greenpoint sit within the borough of Brooklyn (other boroughs being Manhattan, Queens, Bronx and Staten Island). Over 140 books have been written about Brooklyn according to Brooklyn's Borough Controller. Some of the notable Bushwick people over time contained in many of these books have included: Jackie Gleason (1916–1987), actor; Rick Gonzalez (born 1979), actor; Kenneth McMillan (1932–1989), actor; Eddie Murphy (born 1961), comedian, actor; Charlie Murphy (born 1959), comedian, actor; Rosie Perez (born 1964), actress Vincent Schiavelli (1948–2005), actor/writer; Connie Stevens (born 1938), actress; Mae West (1893–1980), actress; Dondre Whitfield (born 1969), actor; Jules de Balincourt; Ryan J. Davis (1982-), theatre and Social media director; Henry Matyjewicz, artist; André' Pierre Charles, graffiti artist; Joell Ortiz, rapper (Slaughterhouse); Da Beatminerz, hip hop production team; D-Stroy, Arsonists (rap group); Jason, Tina and Rachel Trachtenburg; Julius La Rosa (born 1930), singer; Harry Nilsson, singer/songwriter; Jeannie Ortega, singer; OC, rapper; Tony Touch, rapper and DJ; Timbo King, rapper; Q-Unique, rapper; Daptone Records, Indie Music Label; Bushwick Bill, rapper; John Francis Hylan (1868–1936), Mayor of New York City; Phil Rizzuto (1917–2007), baseball player and broadcaster.

Looking beyond Bushwick. and further into the arts, there have been other notable people who lived and completed their work in Brooklyn over the last century, including (alphabetical): Woody Allen (film maker), Isaac Asimov (author), Lauren Bacall (actress), Mel Brooks (actor), Eddie Cantor (entertainer), Aaron Copland (composer), Vic Damone (singer), Neil Diamond (singer), Richard Dreyfuss (actor), George Gershwin (composer), Elliot Gould (actor), Woody Guthrie (musician), Moss Hart (playwright), Susan Hayward (actress), Rita Hayworth (actress), Lena Horne (singer), Harry Houdini (magician), Lainie Kazan (actress), Danny Kaye (entertainer), Carole King (singer), Steve Lawrence (singer), Norman Mailer (author), Frank McCourt (author), Arthur Miller (author), Mary Tyler Moore (actress), Zero Mostel (actor), Martha Raye (entertainer), Mickey Rooney (actor, died in April, 2014), Phil Silvers (entertainer), Barbara Stanwyck (actress), Barbara Streisand (entertainer), Gene Tierney (actress), Richard Tucker (singer), Eli Wallach (actor), Walt Whitman (poet), and Shelly Winters (actress).

On the visual side, the Digital Gallery of the NY Public Library (http://digitalgallery.nypl.org) has available early photos of Brooklyn. A selection of these photos, along with photos from the American Institute of *Architects' Guide to New York City* (AiA 1978), illustrate Bushwick's past along Bushwick Ave (Plate 1.12). Buildings include (left to right): Public School 74, Bushwick Ave; David Medical Building (original William Ulmer Resident), Bushwick Ave; Bustav Doerschuck Residents, Bushwick Av; Bushwick Democratic Club (recently used Pentecostal Church); 1020 Bushwick Av (ca. 1885); and Evergreen Cemetery Chapel.

A further look at Bushwick heritage and the entry to Brooklyn from the Manhattan end of the Brooklyn Bridge (to Williamsburg) is outlined in Plate 1.13.

Plate 1.12 Early Photos of Bushwick Ave. *Left* to *right*: Along Bushwick Ave – Public School 74; David Medical Building (original William Ulmer Resident); Bustav Doerschuck Residence; Bushwick Democratic Club (recently used Pentecostal Church; 1020 Bushwick Av (ca. 1885); and Evergreen Cemetery Chapel (Sources: Digital Gallery NY Public Library 2013, *Architects' Guide to New York City* (AiA 1978). Photo Evergreen Cemetery. NYC)

Plate 1.13 Bushwick and Lower Manhattan. *Left* to *Right*: Hart St (Bushwick Ave); Gates Ave (Bushwick Ave) apartments; church at Menahan St and Bushwick Ave; Knickerbocker Ave 1940s; Fulton Ferry and South St, Manhattan, with Brooklyn Bridge; entry to Brooklyn Bridge, Manhattan (Source: Digital Gallery NY Public Library 2013; *Architects' Guide to New York City* (AiA 1978); and, Old New York in Early Photographs (Mary Black 1976))

Here (from left to right) is Hart St (at Bushwick Ave), residential building with shops below; Gates Ave (at Bushwick Ave) apartments complex with an inner shared courtyard/garden; church at Menahan St and Bushwick Ave; Knickerbocker Ave commercial area, 1940s; Fulton Ferry and South St, Manhattan, with Brooklyn Bridge in background; and Manhattan end of Brooklyn Bridge.

Bushwick has always had a large range of housing, The variety has resulted from building to service an immigrant population from far parts of the world, including business people wanting larger homes. Illustrated here (Plate 1.14) are examples of that varied housing, (left to right) as follows: decorative Weirfield St two story; four story brownstones; apartment complexes; three story walk ups; restored historic stately apartments (Bushwick Ave); and, a three story walk up (note the recent painted themed mural reflecting the Spanish settlement in the area from the 1960s onwards).

A final look at early Bushwick focuses on the early industrial, commercial and professional interests. A grouping of the buildings, particularly representing the German influence, is illustrated in Plate 1.15 (left to right): Ulmer Brewery in West Bushwick (in the vicinity of the Rheingold Renewal Project (examined in Chap. 5); Engine Company 237 Morgan Av represents one of the most important services (fire protection) in Bushwick given the neighborhood's high number of timber houses; and, a further essential service, the Police Department (20th Precinct at Wilson Ave at Dekalb Ave).

Looking further at Plate 1.15 (left to right) is the Central Avenue BMT Myrtle-Wyckoff Avenue elevated railway line. Residents' access to jobs meant a dependency on this rail system. New York City has one of the most intricate and early developed rail systems in the world. The pride of craftsmanship in railway building is illustrated in the tile at the Bushwick Ave Aberdeen St Rail Station. The most famous symbol (final photo) of elevated railway transport in Bushwick (and throughout Brooklyn) is 'the el' as illustrated here running along Broadway, connecting Bushwick with Williamsburg.

1.7 Churches of Bushwick

A sample of Bushwick's historic public churches is outlined in Plate 1.16. The church provided (and continue to provide) an anchor to civil life in Bushwick, serving spiritual and social needs of the community still today. Most immigrant groups in New York City built their religious buildings soon after settling. Here we see (left to right): (1) Mark's Lutheran Church & School; (2) Calvary and Cyprian; (3) St Barbara's Roman Catholic Church; (4) Evergreen Baptist; (5) South Buswick RDC; and, (6) Throop Flushing Church.

The most important commercial street in urban Bushwick was Knickerbocker Ave. Plate 1.17 shows the this avenue in the early 1900s (left to right). This shopping area has been central to the life of generations of Bushwick families. Note at

Plate 1.14 Historic Bushwick housing examples. *Left* to *Right*: Weirfield Two Story; Four Story Brownstones, Bushwick near Suydam St; Meserole Graham Apt. Complex; Jefferson St Three Story Walk Up (Fire Hydrant Open for Cooling in Summer); Bushwick Av Restored Apartments; and, Central Ave Walk Up (Latin American Themed Mural) (Sources: Wikimedia and Brooklyn Historical Society)

the time of this photo there were no cars present. Non-industrial transport then was dependent on horse and cab. The major shops (such as Bekowitz Shoes and Irving Hats) (right in photo) served the people for generations.

This street's importance is highlighted when the range of businesses is examined in Plate 1.18. Note the large number of bar and grills, grocers, and movie theaters as listed.

Plate 1.15 Bushwick Commercial Buildings and Services. *Left* to *Right*: Ulmer Brewery; Engine 237 Morgan Av; Police Dept. 20th Pct. Wilson St.; Central Av BMT Myrtle+Wyckoff Avs; Bushwick Av, Aberdeen St Rail Stn. Tile; The El Broadway, Bushwick and Williamsburg (Sources: Wikimedia Commons and Brooklyn Historical Society)

Plate 1.16 Churches of Bushwick. *Left* to *Right*: Mark's Lutheran, Bushwick Av; Calvary and Cyprian, Bushwick Av; St Barbara's Catholic, Central Av; Evergreen Baptist, Woodbine; RDC South Buswick; Throop Flushing Church (Sources: Wikimedia Commons)

Plate 1.17 Knickerbocker Ave in the early 1900s. Looking from *left* to *right*: note the intensity of footpath shopping; sparse horse and cab street movements; and famous shops such as Berkowitz shoes and Irving hats (Source: Ridgewood Times, July 18 2002)

Perhaps most important were the bakeries scattered over Bushwick, including: Caputo, Circo, Falcetta, Palermo, Montalbano, Saieva, Serro, Stallone, Gaingrasso and Venezia.

On a local commercial note here is a recent quote (author unknown) noting the famous New York City pretzel, sold on many corners of Knickerbocker Ave. The pretzel (origin is German) is ever existent in all parts of Brooklyn, a reflection of traditional preference for a wholesome food.

> The most lasting memory of Bushwick could well be of the soft pretzel, someone always selling these on a street corner. Your equipment was a round fruit basket lined with clean towels, layers of pretzels, and a stick long enough to hold a dozen pretzels wedged on the edge of the basket. These pretzels would sell for a nickel for one and three for a dime. In front of a local movie theatre would mean a hundred pretzels sold.

The pretzel over these years complimented the ever present music created and consumed by the immigrant population. During the 1950s/1960s, for example, walking Bushwick would always mean hearing popular radio tunes. Well known singers then included: Frank Sinatra, Helen Forrest, Connie Francis, Tony Bennett, Johnny Mathis, Elvis, Buddy Holly, Frankie Avalon and Dion and the Belmonts.

Other parts of Bushwick also provided a good number of services complimenting Knickerbocker Ave, including popular services provided at corner shops and on street corners (hot dogs, knishes, chestnuts, fruit, peanuts, etc.). These services gave immediate access to local residents for all their essential needs. Many families worked on an extended credit at corner shops, paid weekly.

Knickerbocker Ave	**Companies**	**Bakeries cont.**
Adam Hat store off Dekalb	Kirsch Soda Stockholm St	Palermo & Canepa Pastry
Al Stone Appliances	Knitting Mill, Stockholm St	Saleva's Pastry
Bee's Womans Shop	2nd Knitting Mill, Dekalb Ave	Serro's Pastry
Bella's Millinery	Rheingold Brewery	Venezia's Bread
Betesh Children's Clothing	Sauerkraut Factory, Wilson/Stockholm	Webers German Bakery
Boy's and Men's Shop	Schaefer Brewery	**Grocers**
Brautmann's Pawn Shop	Sewing Machine, Wilson	Amplo Bros., Wilson
Buster Brown Shoes		Brocia's, Wilson
Charlene's Clothing Store	**Candy Stores**	Dominick's, Wilson
Circo Bakery	Sanzio, Central Av	Jacks, Wilson
Dilberts Supermarket	Betty's, Central Av	Mr Rina, Dekalb
Eddie Cantors	Ann and Harry, Central Av	Mr Jimmy, Central
Fish Store	Frank, Wilson	Myers, Myrtle
Fruit Stores	Amendolora's Soda	Regina, Wilson
Gluckoff Carpet Store		Tony's, Wilson
Harrico Drug Store	**Shoe Maker's**	
Italian Deli corner Dekalb	Morreale, Wilson/Myrtle	**Movie Theatre's**
Izzies Grocery – next to Stella Doro	Spadafora's	Lowes, Broadway
Jack's Deli	**Drug Stores**	Lowes, Gates
King Solomon	Passalaqua's, Wilson	RKO, Madison
Kitchen Set Store	Rabin, Myrtle	Majestic
Linoleum Store		Parthenon
Lofts Candies	**Bar & Grill's**	Ridgewood
Mazzola Furniture	Alps Tavern and Angelo's	Rivoli
Meatland	B&G	RKO Bushwick
Miles Shoe Store	Café Royal	Starr
Moe's Record Store	Enchanted Hour	Wagner
Morris Toyland	Friendly Tavern	Willoughby
National Shoe Store	Grove land Gardens	Wyckoff
Rainbow Shops	Jay's	
Rose shop	La Femina (Tippy's)	**Other Business'**
Scaturro's Supermarket	Piccadilly Inn	Arion Manor
Schotenfields Carpet Store	Royal Cafe	Clam House
Schumacher's Ice Cream Parlor	Saxon	Cypress Gardens
Schwabben Hall	Tip Top Inn	Dekalb Diner
Stallone's Italian Pastry	Willow Greene	Frank & Sophie rose
Stella Doro Bakery		Gasper Polizzi's
Snails and Dried Codfish	**Bakeries**	Gottlieb Deli
Chinese Restaurant upstairs	Caputo Pastry	Paul's Restaurant
Tom McCann	Circo Pastry	Pizzeria, Willoughby/Central
Vinnies Pork Store	Giangrasso's Bread	Rathskeller
Woolworths	Irving & Dekalb Bakery	Ridgewood Terrace
Young folks Store	Montalbano's Pastry	Ventimiglia's Restaurant Victory Hall

Plate 1.18 Businesses of Knickerbocker Ave and Surrounds, 1950s (Source: City of New York Archives 2013a, b)

Brooklyn life was disrupted a number of times in the twentieth century. This included World War 1, with Brooklyn providing a major war input via its effort through it's industrial base. The Great Depression of the early 1930s hit Bushwick families as hard as any in America. There was a scarcity of jobs for the working class over all of New York City. The onslaught of World War 2 again saw Brooklyn rising to the industrial needs of the nation. During these crises years Brooklyn residents relied on its working class strengths and historical industrial background. Northeast Brooklyn had significant factories; and, access to the East River inlets to ensure easy water transport. Bushwick residents were a short bike ride or public transport trip to Northeast Brooklyn industries, such as the Brooklyn Naval Yard (Greenpoint) and Domino Sugar (Williamsburg). By the mid-1940s, with WW2 over, life returned to normal and the building of a nation was on the agenda again. Dwight David 'Ike' Eisenhower, World War 2 Supreme Allied Commander, was in the Whitehouse and the United Nations (located in Manhattan) had been established to progress peace.

In Bushwick children relied upon easy access from homes into safe streets, with eyes of neighbors always present. Rauscher (author) recalls key places for Bushwick youth in the Dekalb/Central/Wilson/Knickerbocker area (post 1940s), including: bakeries on Wilson Ave and at Knickerbocker Ave (Stella Dioro); police station of 83rd Precinct (sponsoring the Police Citizens Youth Club of Bushwick); Knickerbocker Park (now named Maria Hernandez Park); movie theaters wide choice of Starr, Willoughby, Rivoli, and Loew's of Broadway; Woolworth's department store; Hymies in Dekalb Av, one of many candy stores in walking distance from homes; primary schools 68 and 74; Bushwick Library (Dekalb branch) in Bushwick Ave; and, the Pigeon Store that served many hobbyists on roof tops of Bushwick (flocks in formations often seen circling the skies daily).

Throughout Bushwick's commercial history there were always businesses in Flushing Ave and Morgan Ave, as well as Knickerbocker Ave. A secondary business center also existed in the Williamsburg area (Kent Av and the Williamsburg waterfront). Companies in that district included: Rheingold and Schaefer breweries; Welbuilt Stoves; Phelps Dodge; Williamsburg Steel; Domino Sugar; and, the Brooklyn Navy Yard (as noted earlier). The Navy Yard was owned by the Federal government, with the City of New York purchasing it in the late 1960s. Currently the site is being turned into industrial park, residential and open space uses.

Bushwick has a rich history in streetcars. The first elevated railway in Brooklyn was here in the Lexington Ave Elevated (1885) (Gates Ave and Broadway terminus). This line and others serving Bushwick gave easy access to downtown Brooklyn and Manhattan (leading to rapid residential development of Bushwick) as noted earlier. The success of industry in Brooklyn and Bushwick meant the more affluent were soon to commission mansions along Bushwick and Irving Aves in the early 1900s. Wikipedia notes that:

> Start: taken from web (13 Nov 2013).
> Styles included Italianate, Neo Greco, Romanesques Revival and Queen Anne. In this tradition Bushwick became a center of culture with playhouses and the nation's first theater with electric lighting. By 1950 the area contained one of Brooklyn's largest Italian American neighborhoods.

Bushwick however continued to face decline in the 1960s (including blight and poverty).

Wikipedia goes further in outlining the white flight out of Bushwick and the area's resulting economic depression. There is, for example, a description of the blackout resulting in riots, fires and looting on the night of 13 July 1977, which is discussed in the next chapter.

1.8 Chapter Summary and Conclusions

This chapter looked at Bushwick's early history, from the indigenous community of North American Indians to European occupation. The significance of the independence movement was explored as was the lead up to the establishment of Bushwick Town, one of several Old Brooklyn towns (pre-1850). The emergence of a settled community of Bushwick was illustrated in looking at the early churches and public services in Bushwick. Finally, Bushwick's urban life in the 1940s–1960s was examined. With this understanding of Bushwick's historical background, in particular its urban development, the reader was introduced to life in Bushwick in the 1970s to current times, including the low point of the arson fires of 1977 (noted earlier).

References

American Automobile Association (2000) 9L4 Street Map. American Automobile Association, New York City

American Institute of Architects (1978) Guide to New York City. American Institute of Architects, New York City

Brooklyn Historical Society (2000) History of Williamsburg. Brooklyn Historical Society, New York City

City of NY Archives (2013a) Old Brooklyn's Towns Pre-1854. City of New York, New York City

City of NY Archives (2013b) City of Brooklyn emerges in the late 1850s. City of New York, New York City

Dept of the Interior (Stiles) (2013) The Battle of Long Island 27 Aug 1776. Dept of the Interior (Stiles), Washington, DC

NYC Public Library (2013) Digital Gallery (Early Photos of Bushwick Av.). NYC Public Library, New York City

Richmond Hill Historical Society (2013) Map of Long Island by the Dutch in 1635 Depicting Native American Tribes. Richmond Hill Historical Society, New York City

Ridgewood Times (2002, 18 July) Knickerbocker Ave in the early 1900s. New York

Chapter 2
Bushwick Planning: 1970s to Current Times

2.1 Chapter Introduction

This chapter examines urban life in Bushwick in the 1970s to current times. It includes the demise of Bushwick in 1977 and the recovery in the 1980s to current times. The reader is introduced to the recovery up close in accounts of Rauscher (author) which were gained from field trips he made to Bushwick, Northeast Brooklyn and other parts of Brooklyn from 1979 to 2010. Finally, the movement to green designs and their importance to Bushwick's future development is reviewed. The chapter ends on the note of opportunities for Bushwick to embrace sustainable urban planning (SUP) approaches.

2.2 Bushwick: 1970s and Demise

Bushwick's boundaries (Plate 2.8) today is defined (matching the historical boundaries) by Brooklyn Community Board 4 (Flushing Avenue on the north, Broadway on the southwest, the Queens Borough line to the northeast and the Cemetery of the Evergreens on the southeast). By the mid -1950s, migrants began settling into central Bushwick. The U.S. Census records show that Bushwick's population was almost 90 % white in 1960, but dropped to less than 40 % in 1970 (U.S. Census). As white families moved out of Bushwick (noted in the previous chapter), new immigrants took over homes in the southeastern edge of the neighborhood, closest to Eastern Parkway. A strong desire among these new residents towards home ownership and block associations helped the neighborhood survive the economic and social distress of the 1970s.

> The signals of trouble in Bushwick were evident in the report *Preventive Renewal Areas, NYC* (NYC City Planning Commission 1972). This report mapped a large number of 'preventive renewal areas' (Plate 2.1).

R.C. Rauscher and S. Momtaz, *Brooklyn's Bushwick - Urban Renewal in New York, USA: Community, Planning and Sustainable Environments*, DOI 10.1007/978-3-319-05762-0_2,

Plate 2.1 Preventive renewal areas, New York City 1972 (Source: NYC City Planning Commission 1972 (NYC 1972))

A closer look at the extent that Bushwick and its surrounding were signaled as needing attention is provided in Plate 2.2. Bushwick is central-right in the picture with the designated 'preventative renewal area' extending north along Myrtle Ave into Queens. Note also the targeted renewal areas included parts of Greenpoint and

Plate 2.2 Bushwick within preventive renewal areas 1972. Bushwick (*center-right*) is shown with a large section labeled as 'preventative renewal area' (Source: NYC City Planning Commission 1972 (NYC 1972))

Williamsburg (the two blackened areas to the left of Bushwick along the East River, respectively). The report's comment on Bushwick was:

> Bushwick has one of the highest fire rates in the city. Repair loans were meant to remove housing violations (and contain the spread of fires) and help preserve the neighborhood. If nothing is done to combat the deteriorating housing stock this area will be lost as a viable residential area. Failure to act would be disastrous.

The conclusion of the report was prescient for what was ahead for Bushwick residents, leading up to the fires of 1977. The City explained to the Federal Government in the early 1970s (after the above report) that a multi-billion dollar loan was required to get the city on a financial pegging to carry out required programs. The request was a major news item across America, and the Federal Government refused to assist.

At the same time, the past history of the building of high-rise public housing blocks was being shelved in favor of low-rise medium density developments. Plate 2.3 shows the extent, at one stage in New York City history (1960s/1970s), of the policy to build high rise housing blocks. The area is in a redevelopment area of New York City. It was apparent that the social isolation of living in high-rise buildings in poorer areas had contributed to administrations going back to low and medium rise housing in the 1980s onwards.

Plate 2.3 High rise in a redevelopment area in New York City (Source: City of New York Archives 2010)

In contrast to high rise housing promoted by administrations in the 1960s and 1970s as the answer to mass housing, the housing tradition of low rise in most of early Brooklyn (Chap. 1) was given a lower priority. The extent that early Brooklyn embraced low rise for many years, and still achieving a high density is illustrated in the 'rows of housing' of three story walkups, mixed with medium rise, in the Brooklyn neighborhood of Sunset Park (Plate 2.4). The lack of open space however, even here, was a shortcoming of Brooklyn over many years (in spite of providing a range of 'city playgrounds' and opening school playgrounds for local residents in the long summer school holidays of 3 months).

The blackout and the fires in Bushwick in 1977 were devastating. It appeared there had not been sufficient advancement programs implemented pre-1977, these perhaps preventing the severity of the incident. A discussion on this question is presented later in this chapter, examining author (Rauscher) research in Bushwick. In addition the writings about Bushwick by urban planners and others is commented on.

2.3 Recovery and Urban Planning

In 1979 swaths of Bushwick's more blighted areas (which were similar to sections of South Bronx troubled areas in the early 1970s) resembled bombed out European cities of World War 2. It was obvious the city was not coping with urban problems

Plate 2.4 Low rise residential at Sunset Park, Brooklyn (Source: City of New York Archives 2010)

at the time and preventative measures had not been listened to in the years prior. There were signs, however, that citizen and company investment in some areas of Brooklyn, including Bushwick, was opening up. These areas included: Williamsburg, Greenpoint, Brooklyn Heights and the Downtown Brooklyn (looked at in Chap. 5). The restoration of homes was becoming popular.

Plate 2.5 Up From Flames (Source: Brooklyn Historical Society, Archives 2013)

Mapping Bushwick's recovery from 1977 to present was undertaken by the *Up From Flames* group (Plate 2.5).

The group's website (www.brooklynhistoricalsociety.org) summarizes the work of the group including: interactive history map of Bushwick completed by the Academy of Urban Planning students; a Bushwick 1977 landscapes slide show; a Bushwick Today; and Points of View. Community planning in NYC, commencing in the 1990s, was considered the best means of gaining the best outcomes for neighborhoods and the city overall. On looking at the history of planning affecting Bushwick, Up From Flames comments: "A policy of planner shrinkage allowed Bushwick to sink into ruin. Cutbacks in social services, including fire protection, made the neighborhood vulnerable to fire. The nation witnessed the desperation of this abandoned community through the rioting and looting that occurred during the 1977 New York City Blackout". On a positive historical note, Up From Flames goes further: Once the critical needs of Bushwick were recognized, the media put pressure on the City to take action. Bushwick could no longer be ignored. Under Mayor Koch, collaborative planning between city and local government created innovative long term solutions to Bushwick's housing crisis. Finally, Up From Flames reinforced a message picked up by the urban planners: "Today's Bushwick is the product of carefully considered public policy that laid the groundwork for growth and private investment. The current challenge for policy makers is to sustain affordable housing in todays heated real estate market."

The rise of Bushwick from the late 1990s and through the 2000s was the signal of renewal residents had worked for since the 1977 fires. A walking tour of Bushwick www.nyc.gov/html/hpd/html/about/bushwick-walking-tour.shtml provides an overview of the renewal investment made by the City of NY and not-for-profit organizations in the 1990s. One such not-for-profit group in Bushwick that has done considerable work for renewal is the Ridgewood-Bushwick Senior Citizens Corp (RBSCC) www.rbscc.org, working tirelessly from the 1970s to today and was involved in several of the projects.

Major projects completed by the RBSCC during (and noted in the walking tour program) included Buena Vida Nursing Home, 48 Cedar Street, corner Evergreen Avenue (was opened in 2001). This home (eight story and containing 240 beds) caters for senior citizens and came about through the efforts of the Ridgewood Bushwick Senior Citizens Council working with the city, state and federal governments. Partnership New Homes Program, 55–67 Cedar Street, consists of eight two-family homes. This was an initiative of the New York City Department of Housing Preservation and Development and the New York City Housing Partnership. The brick homes, feature a three-bedroom owner's unit with a full basement and fenced backyard, and a one-bedroom rental unit. A total of 149 two-family homes were the initial start of this program, with homes scattered throughout Bushwick.

Central Avenue contains additional Partnership homes, including: low-income housing for seniors (143 Himrod Street); The Ridgewood Bushwick Senior Citizens Council 60-unit building, owned and managed by the Council. Adjacent to the senior housing is the New York City Housing Authority's (NYCHA) Hope Gardens. This development is located on a number of blocks throughout Bushwick (e.g. 140 Himrod Street) and consists of three-story townhouse-like structures. More than 1,000 units of NYCHA housing were built in several phases in the early 1980s as part of an Urban Renewal plan designed to redevelop the most devastated areas of Bushwick.

Two multi-family buildings that front both sides of Harman Street (160 and 173) were built in 1995 as 42-unit low-income rental developments under the Permanent Housing for Homeless Families Program (City/State "85/85" Program) (sponsored by The RBSCC). Other homes in the vicinity of Himrod St were rehabilitated through the Neighborhood Entrepreneurs Program. In a research field trip in 1990 there were signs of renewal not only in Bushwick but in most neighborhoods in Brooklyn.

An overlook view of life in Bushwick was provided by Village Voice in 2002 (*Village Voice* 2002). Summarizing statistics contained in the article illustrates the high number of working class people who still predominated in the area at the time. The article notes that Bushwick was once filled with textile factories, as well as breweries. Many of the new residents were Hispanic, with a high dependence on welfare benefits. Bushwick was still infamous as one of the poorest sections of Brooklyn at the time (still suffering the stigma of the fires of 1977). This is to be examined more closely in Chap. 5. The Village Voice states that up to 40 % of the population was reliant on public assistance in 2002. The Village Voice also noted a turnaround in housing was well underway, seeing empty lots being developed into city-subsidized housing. Finally, the article notes that many houses were city owned properties, sold or rented through the Department of Housing, Preservation and Development (DHPD).

A concise history of changes in Bushwick before and after the fires of 1977 is presented by Steven Malanga (Make the Road 2008) (www.maketheroad.org). His focus is Bushwick as a Brooklyn neighborhood that is gradually recovering from decades of misguided urban policies. Malanga takes a tour of Bushwick through the eyes of a local resident as follows.

Start: taken from web (13 Nov 2013)

These days, when Morris Todash walks the streets of Bushwick, a two-square-mile neighborhood of 100,000 people in central Brooklyn, he likes what he sees. On the long-abandoned seven-acre site of the former Rheingold Brewery, new two-family homes and condominiums have sprung up. On the side streets along Broadway—not so long ago, pockmarked with desolate lots where stray dogs wandered amid burned-out cars—more new homes arise and old ones get impressive face-lifts.

New businesses—an organic grocery store, a fashionable restaurant—seem to be opening on every corner. Todash, whose insurance firm has served the neighborhood for more than 40 years, can hardly believe that this is the same Bushwick that became synonymous with urban chaos during the late 1960s and early 1970s, ravaged by fires, rioting, and looting until it resembled a war zone. "When I first came here to open a business, this was a shopping destination for all of Brooklyn," Todash says of the neighborhood's commercial district. 'After the looting, no one wanted to come here any more.'

Ruinous policies battered it down. So total was the devastation that even as New York began rebounding in the mid-1990s, Bushwick remained largely untouched by gentrification. Only recently—after years of tireless work by government (especially the police), local groups, and the private sector—has the revitalization of this once-proud neighborhood begun. With Bushwick beginning to thrive again, New York City has finally left behind the disorder and failure that flowed from the misguided reforms of the sixties and seventies. Yet if Bushwick is back, no one should forget what happened to it.

End: taken from web (13 Nov 2012)

Jerilyn Perine, an urban planner, led the Department of Housing Preservation and Development between 2000 and 2004, comments on the urban health of Bushwick in a paper at the International Cities and Towns Society conference in Australia (Perine 2005). The paper examined a brownfield site (abandoned urban lands, often contaminated from previous uses). The site was the Rheingold Brewery (reviewed in Chap. 5). It was Perine's comments however that contribute further in understanding the urban demise of Bushwick in the 1970s. Perine notes that the community was "wary of government schemes to improve their community. Often such efforts resulted in wide spread demolition and displacement of long-term residents." She notes that Bushwick residents had recently added "gentrification to their long list of worries and concerns."

Perine reflects on why the 1970s was not a good time for America's cities. She explains: "The stage for their (cities) decline had been set in the decades following World War 2 when the Federal Government, seeking to address critical housing shortages, began to subsidize suburban development:" This development was done through highway construction and low cost, long term mortgage lending. Perine points out that "the exodus of stable, moderate and middle income urban American families had begun taking their local taxes with them. In the 1970s high inflation and unemployment followed the decline of manufacturing. The withdrawal of the Federal Government from housing and other urban assistance programs left most American cities reeling from population losses, physical decline, financial instability and rising crime rates."

Perine, in reviewing further the impact of all the trends noted above, noted that the loss of manufacturing jobs (1/2 million within 20 years in New York City in 1960s/1970s), cut off the supply of jobs to workers with little education. She writes that "such jobs had been the stable of waves of immigrants coming to New York and enabled them to move up the economic ladder. Now immigrants on the bottom would stay there, with few opportunities for jobs." She points out that the exodus of jobs left thousands of acres of vacant industrially zoned land, thus impacting the very same neighborhoods that the neediest families lived in. Perine notes that moderate and middle income families moved out to outer suburban areas and Long Island, (while) neighborhoods like Bushwick attracted poorer families. She writes: "With unscrupulous landlords filling their properties with families on public assistance (and banks reluctant to make loans in such areas) the housing conditions deteriorated. Given buildings simply became worth more to their owners if they were burnt down in a fire and insurance could be collected, then if they were maintained for rental housing." Perine presents the alarming statistic that "between 1970–1981 over 321,000 housing units were lost to the housing market (in New York City) through fire (primarily arson), deterioration, abandonment and demolition." She goes further: "This calamity, unprecedented in America's history, was not evenly distributed across the City, but rather it was concentrated in three primary areas: … South Bronx, Harlem, and four communities in Brooklyn, including Bushwick." Perine quotes Robert Caro's book The *Power Broker* (Caro 1974). Caro outlines in his book the New York City housing catastrophe at the time, particularly in the South Bronx (the book being a definitive work on the decline of New York City's neighborhoods).

Perine is critical of the Government's lack of intervention in places like Bushwick in the 1970s and when actions did take place they were poorly conceived and executed. She states: "Whole blocks were demolished, displacing long time residents and leaving in its wake empty lots as redevelopment schemes could not get off the ground." She continues: "With its extreme poverty (by the 1980s, 80 % of Bushwick's adults were unemployed), and low scale mostly wood frame buildings, Bushwick's housing stock was vulnerable to arson. Entire blocks simple went up in flames." She then addresses the event on a hot summer night in 13 July 1977 (9:30 pm) when the city suffered a major blackout. In Bushwick, she writes (its worth reflecting on the Bushwick cultural factors outlined above): "looting came in three district waves. First and almost immediately were the career criminals, in abundance in Bushwick at that time…They descended on Broadway, Bushwick's main commercial street, with an immediacy and seriousness of purpose that overwhelmed the local police precinct. Second came the 'alienated teenagers' and third, poor people exploiting the lawlessness and greed that the situation presented."

Perine continues, in detailing this urban lesson for Governments to consider in the future: "While 31 low income neighborhoods (in New York City) were damaged in looting that night, Bushwick suffered the worst devastation…crowds began to burn buildings as well. By the time the lights came back on the next day 134 shops on the main retail street, Broadway, were looted and damaged and 45 of them were burned out and destroyed. By the time the 1970s came to an end (two and half years after the arson fires of 1977) Bushwick had lost 20 % of its housing, one out of every five apartments was destroyed; one third of its population left and half of its businesses were lost."

Perine points out, on the positive response to the ills of the City, that the City launched the Ten Year Plan (1986), with the plan continuing at the time of Perine's paper (2005). She notes that since 1986 more than 200,000 housing units had been rehabilitated or newly constructed through different City programs. She notes: "Thousands of units were rehabilitated or newly built in Bushwick during this period. Stores began to return and new waves of immigrants continue to move in."

In addition to Perine's insights, an on-the-ground examination of Bushwick was undertaken by John A Dereszewski (self published 2007). He comments on Bushwick of the 1970s as well as reviewing the recovery of Bushwick up to 2008. His work compliments the writings explored above, namely, Up In Flames (www.brooklynhistoricalsociety.org) (2013), Malanga (Make the Road 2008), and Perine (2005). Derezewski notes that arson and abandonment devastated Bushwick's central core well before the black of 1977. He comments that "on blocks like Himrod St and Greene Ave (between Central and Wilson Aves) every building had been abandoned. Due to funding shortages…most of the abandoned buildings, almost all of which were unsalvageable, were not demolished. This created the impression of essentially living in a war zone." He indicates, that before the blackout, several Government actions were taken that strengthened the community. In the early 1970s the City opened a Neighborhood Preservation Office charged with completing a locally based development plan. This office took the initiative to develop this plan in close collaboration with local community board, Brooklyn CB4. The community board was able to open a local office and hire a small staff in May 1977. In spite of these positive moves, the arson fire (Derezewski writes) was an entirely Bushwick event. He writes that it added an exclamation point to Bushwick's particular plight and brought the full spectrum of the community's decade long descent into arson assisted housing abandonment and absolute despair to the general public's consciousness.

2.4 Recovery Up Close

Dereszewski (unpublished 2007), commenting above about Bushwick in the 1970s comments on the first steps to recovery. He comments on a Bushwick Action Plan (as promoted by the CB4) with the community insisting low-rise scale that typified Bushwick be maintained. Derezewski comments: "…housing would only be two or, at most, three stories high and would be constructed on existing block fronts. In all, the City would construct 1,076 low income and 243 senior citizen housing units in Bushwick during the early to mid-1980s. There would be no additional 'super blocks' in Bushwick."

Derezewski comments on the introduction of the NYC Housing Partnership, a collaborative effort involving the City and the business community. He notes this introduced two family housing that (through subsidized mortgage rate and tax abatements, was affordable to most working families). Derezewski points out that Partnership Housing became the dominant form of new residential development throughout all portions of Bushwick. He writes: "Consistent with the Action Plan, these residences preserved the exciting community scale and, being constructed on vacant land, did not displace existing residents." Derezewski focuses on strategies that needed immediate and concrete results, while the area waited for new housing, particularly to the long-term residents of Bushwick who had withstood the worst. He notes the actions fo the City to build a sense of community in outlining programs such as: demolition (dangerous buildings eliminated); tree planting (to address the dearth of street trees with the City responding by planting several thousand trees by the late 1970s); parkland development (the City identified large vacant areas for recreational development); economic development (noting the strength of the Knickerbocker Ave shopping district); a 'Bushwick Initiative' effort to stabilize and improve housing, combat crime and improve the health and quality of life in a targeted area commenced; private market housing, at an acceleration space, had been constructed in Bushwick (many are three story 'Fedders Housing' units); and, finally continued migration of young artists and professionals from Manhattan to Williamsburg, went further to Bushwick along the 'L' rail line.

Derezewski writes further: "this migration trend began to transform the formally depressed industrial zone along Bushwick's northern border into a gentrifying district of converted lofts, coffee houses and exciting bars and restaurants. It has also steeply increased the property values and rental costs in this previously low income community." He concludes that special attention should be given to preservation of Bushwick's most stable and architecturally significant neighborhoods. In a sense this takes the story back to the original foundations of Bushwick as outlined in Chap. 1.

In every story there is always an unsung hero, perhaps in the Bushwick story of the 1970s that recognition would go to Father John Polis (a Monsignor) (born in New York City during the Depression), St Barbara's Roman Catholic Church, Central Ave, Bushwick (Plate 2.6). The author (Rauscher) interviewed Father at his church in 1982 to discuss the recovery of Bushwick (see also the role of churches in Bushwick in Chap. 1). In writing about Father Polis, Chas Sisk of the Graduate School of Journalism, Columbia University (Columbia University) (12 Dec 2003) wrote of Father's significant role in supporting the community of Bushwick in trying times. Sisk writes: "he worked tirelessly on key issues of housing and community services". Sisk adds "that people saw Father as a 'radical' ahead of his times in responding to the needs of Bushwick residents and taking up the cause with city hall. He worked closely with young people at the office of El Puente, a youth center located next to St. Barbara's. He was aware that two-thirds of Bushwick's population was Latino with a per capita income of half the average. Polis's hero was Ivan Illich, a Catholic theologian active in the 1960s and early 1970s. Illich became a popular writer on how to combat poverty and injustice. Father Polis co-founded in

Plate 2.6 Father Polis, Parish Priest St Barbara's R. C. Church, Bushwick (Source: Raymond Rauscher)

1978 the East Brooklyn congregations to take up social causes (see Chap. 6 to learn about the innovative high school that this group founded in Brooklyn)."

While home ownership during the recovery (and still today) was less than 20 % there has always been an owner pride of ownership (U.S. Census 2007). On two of several field trips to Bushwick post -1977 fires, by the author (Rauscher) the recovery from the demise was noticeable (Plate 2.7). Here the securing of housing is seen in three streets in the Dekalb/Central Ave section of Bushwick. From left to right, the first two photos show two-story family homes in Cedar St (note the added security in the second house). Basements were often rented out. The next two photos (Cedar St) show a single story home (set back from the street) under restoration in 1979 and completed by 1982. The pride in home ownership is convincing here. The last two photos show new two-story town houses in Menahan St and three story walk up apartments in Central Ave.

2.5 Community Recovery

While housing was shown signs of recovery as noted statistics on Bushwick reflect the continued social and economic needs of the area. The population of Bushwick was 129,980 in 2007 (U.S. Census), with about 1/3rd of residents born overseas. At that time the population falling below the poverty line was still one in three. Nearly three out of four residents in 2007 came from the Hispanic-American community, a trend that had started in the 1960s as noted in Chap. 1. Finally, In spite of many

Plate 2.7 Renewal in Bushwick neighborhoods. *Left* to *Right*: Two Family Homes (Cedar St); Single Family Homes, with added security (Cedar St); Single Street Set Back Home under restoration in 1979 (Cedar St); Completed restoration of same home in 1982; new town houses (Menahan St); three story walk up complexes (Central Av) (Source: Ray C. Rauscher)

Plate 2.8 Bushwick Community Profile. *Left* to *Right*: Maria Hernandez Park; Knickerbocker Av; BMT Canarese Line at Montrose; Irving Square Park; Myrtle Ave Line; and Bushwick Public Library (Source: Wikimedia)

advances to curtail crime in Bushwick, post the 1977 fires, the area in 2007 still had one of the City's highest rate of felony crimes (U.S. Census).

A look at the community side of the recovery can be seen moving around Bushwick (Plate 2.8). Here (left to right) is the upgraded Maria Hernandez Park;

Knickerbocker Ave; BMT Canarese Line at Montrose; Irving Square Park; Myrtle Ave Line; and Bushwick Public Library.

By the early 2000s Bushwick started to benefit more widely from the private and public interests in renewal. The improvements experienced in Bushwick by 1999 was summarized in an article from *The People's Voice* (a Bushwick newspaper). Another sign of positive improvement was the renovations to St. Barbara's Church as reported on in 2000. Artists had discovered Bushwick by the 1990s as witnessed in the creation of the Arts in Bushwick group in early 2008. The group is completing a cultural vitality project and by 2013 had successfully staged a number of weekend arts festivals with 50+ venues having open house shows. The vigor of community, business and government efforts are reflected in the content of this organization's web site (www.artsinbushwick.org). Finally, Bushwick gentrification and its effects on the community continue today as a real estate boom (early 2000s) removed many homes from the poorer sectors. (as noted in comments by urban planners and others earlier in this chapter).

2.6 Moving to Green Design

The City of New York has a long history of assisting industry and its own administration to embrace green design features in buildings. In 2010 the City launched *The Green Building Handbook* (City of New York 2010) to provide further incentives in this area of sustainability planning. The City applies its green design features under Local Law 86, known as the 'Green Building Law' (City of New York 2005). The law applies to City-owned and City-funded buildings, but is also used as a starting point for any developer interested in green building. In addition green buildings are also a key component of PlaNYC: A greener Greater New York (PlaNYC) (www.nyc.gov/planyc), the City's long-term sustainability plan (to be examined in Chap. 3).

The US Green Building Council's Leadership in Energy and Environmental Design (LEED) rating system is a standard used throughout the country and applied to New York City. This ratings system can also be applied to neighborhood development also (very applicable to the renewal programs to be examined in Chaps. 3 and 4). The Green Building Handbook points out that New York City is more energy efficient than many other American cities given its dense urban fabric. It is also noted in the handbook that, with car ownership low in the City, the City's greenhouse gas emissions is near 80 % from energy consumed in buildings. Reference is made in the manual to the green building educational results stemming from work of the then Brooklyn based (now closed) Brooklyn Center for the Urban Environment (BCUE). The Centre chose to rehabilitate a vacant factory using the LEED standards to create a model for energy and building materials efficiencies (Plate 2.9).

In addition to government and non-profit group's renewal initiatives (e.g. BCUE's example) in green design, private industry has also been busy with applying green

Plate 2.9 Green Design, Brooklyn. Exterior of Green Design Brooklyn Center for the Urban Environment (Source: *The Green Building Handbook* (City of New York 2010))

design to restoration and infill projects. Of particular notice in Bushwick is the integration of post-modern buildings and the inclusion of green design features (i.e. energy saving inclusions). The Troutman Street condos, for example, represent this trend toward green design (Plate 2.10).

A building, representing post modern design, using the LEED standards (buildings are certified under the program) was a further building under Thread PL (Plate 2.11). Tread (unpublished) comments on the need "integrating sustainable principles in every aspect of the design process…there is a need to incorporate sustainability at the outset of the design as it ensures an efficient and well thought out results." At the time, green design groups such as New York State Environmental Research and Development Authority (NYSERDA) (solar and construction efficiencies) www.nyserda.ny.gov and GreenHomeNYC (a not for profit group whose aim is to bring green design information to resident and developers to achieve a 'systemic changes to building practices'). Given the arson fires Bushwick experiences (Chap. 2) the evolution of safer buildings has been on the City of New York administrators' agenda. The City has been running the Environmental Protection Green Building Competition (rewarding excellence in sustainable design and systems integration).

Building green within Bushwick, and through the urban world, can produce additional health benefits. The Green Building Handbook argues that by planting

Plate 2.10 Bushwick Post Modern Building Troutman Street Condos (Source: Thread 2008)

vegetation on roofs will not only minimize storm water runoff and associated pollution to waterways, but also creates a positive attribute for the residents. Plate 2.12 is an illustration of a roof top garden in Manhattan, provided as a model by The City of New York for developers to consider in new or renewal projects.

2.7 Chapter Summary and Conclusions

This chapter looked at urban life in Bushwick in the 1970s to current times. Included was: the demise of Bushwick in the 1970s; the planning of recovery; the recovery steps, including author field trip results; and, the movement to green designs in Bushwick.

The next chapter looks at how Bushwick, Brooklyn and all urban areas could benefit from applying planning and development approaches that incorporates sustainable urban planning (SUP).

Plate 2.11 Leadership in Energy and Environmental Design (LEED) Certified Building (Source: Thread 2008)

Plate 2.12 Roof top gardens and solar installation. Location: Battery Park City, Manhattan (Source: The Green Building Handbook (2010). New York City Economic Development Corporation)

References

Brooklyn Historical Society, Archives (2013) History of Williamsburg. Brooklyn Historical Society, New York City

Caro R (1974) Power broker. Random House, New York City

City of New York (2005) Local Law 86 (Green Building Law). City of New York, New York

City of New York Archives (2010) The Green building handbook. City of New York, New York

Dereszewski J (2007) Bushwick notes: from the 70s to today. Up From Flames c/Brooklyn Historical Society, New York. http://upfromflames.brooklynhistory.org/uff_resources/

Malanga S (2008) The death and life of Bushwick. Spring Publications, New York City

New York City Planning Commission (1972) Preventive renewal areas. The City of New York

Perine J (2005) The Rheingold Brewery's Brownfield redevelopment in Bushwick, Brooklyn, New York. International cities and towns conference, Sydney

The People's Voice (1999) Newspaper. Bushwick, New York City

Thread PL (2008) Architectural projects completed (in company profile). Thread, New York City

U.S. Census (2007) Bureau of Census, U.S. Census, Washington, DC

Chapter 3
Planning for Sustainable Communities

3.1 Introduction

This chapter discusses the increasing importance of sustainable urban planning (SUP) in all urban areas, from Bushwick, Brooklyn or anywhere in the world. The chapter firstly defines sustainability. The defining leads to a review of the movement of urban planning to ecologically sustainable development (ESD), including sustainability criteria. The chapter reviews the historical changes in urban planning schools over more than one half a century. To better understand the principles of ESD the history of its introduction and application the chapter examines three aspects of that evolution: actions at international levels; actions at non-government organizations levels; and, currently applied ESD based planning frameworks. The position of the City of New York (thus affecting Brooklyn, including Bushwick) in embracing the planning for sustainable communities is commented upon. The chapter provides a planning platform for a critique of planning initiatives in Brooklyn, including Bushwick, in the following chapter.

3.2 Early Universality Thinking

The subject of the need for greater cooperation among members of a community (town, city or region) and means to better conserve natural resources arose over a hundred and fifty years ago. The thinking in the 1860s–1880s, in particular, introduced concepts of universal thinking placing nature as central to life. Thus the book has been dedicated to four universal thinking writers (Walk Whitman, Helena Blavatsky, Henry George in the 1860s–1880s; and later in the 1930s/1940s Henry Miller). A summary of this universal thinking and its roots to sustainable thinking as may apply to New York City, Brooklyn, and Bushwick follows.

Universal thinking goes beyond the individual to a wider community commitment to the laws of nature (from local to world community). This thinking is based on

R.C. Rauscher and S. Momtaz, *Brooklyn's Bushwick - Urban Renewal in New York, USA: Community, Planning and Sustainable Environments*, DOI 10.1007/978-3-319-05762-0_3,

greater references to the natural world and conservation principles (as against unlimited resources). This thinking comes through over the years from most of the religions of the world, from the Bible to the Koran.

In the past writers delved into universal thinking placing nature as central to life, including: Coleridge, Carlyle, Plato, Swedenborg, and Bohme. All these authors wrote about the liberating of the individual from oppressive societies. This thinking was advanced in the 1830s, especially at Concord Massachusetts, USA. At that time in America this thinking was a challenge to the older more conservative and traditional European ways. The thinking questioned the conventions of eighteenth century and advocated reform of the state and society. This in turn led to campaigns in areas of: humanitarian causes; democratic aspiration; environmental planning; a more people scale architecture (i.e. Louis Sullivan); and modernism in the arts (i.e. Alfred Stieglitz).

Universal thinking emerged as a world movement in the 1860s–1880s, a time when New York City was developing as a world class city. The thinking reinforced principles of: the innate goodness of man; role of inquiry; and personal experience as a foundation of truths. Those writers embracing these premises included: Emerson, Thoreau, Margaret Fuller, Orestes Bronson, Elizabeth Peabody, James Freeman Clarke, George Ripley, Bronson Alcott, WE Channing, Walt Whitman (note this books several comments and poems written in this text to acknowledge Whitman), Herman Melville and Nathan Hawthorne.

Many groups absorbed aspects of universal thinking within their formal organizations, including: the Theosophical Society (1875) (New York City worldwide launch); Christian Scientists (1879); and, Baha'is (1880). Major Baha'i documents announced by Baha'ullah (born 1817, Persia) were done in New York City, where an Office of Publication and a secretariat at the United Nations operate from today. Wider advancement of universal thinking was overshadowed by global events such as WW1, Great Depression; WW2, Cold War, Vietnam War, Middle East Wars, War on Terrorism, and a multitude of conflicts within states (from Pol Pot in Cambodia to Mugabe in Africa). The contributions of the residents during the first few of these events were commented on in Chap. 1. The twentieth century did however see emerge in the last quarter a revival of universal thinking. This was espoused in movements such as: green conservation; alternative health; and, social justice. New York City residents and industry have often been in the mix within these movements as noted in Chap. 2.

Society entered the year 2000 with these movements finding common threads of principles. For some people there appeared to be centuries of ignorance, including benign or deliberate neglect of natural conservation. Critical issues nearing the end of the first decade of the twenty-first century (2010) and under continual debate in 2013 include: global warming; and, finite resources of water, minerals and oil. Currently the access to knowledge about these issues has been widely spread by social media, internet groups, and new advocacy groups, including: AVAAS (USA), GET UP (Australia), SIMPOL (Simultaneous Policies) (UK), and The Elders. This latter group has membership of Nelson Mandela, Jimmy Carter, Bishop Tutu and other world leaders who focus on solutions to world concerns, including housing people in urban places and environmental protection.

America now has a population of 313.3 million (2012 US Census Bureau) (an increase of 27 m from the 286 million in 2005) (US Census Bureau). This population represents 10 % of the world population, with a consumption percentage considerably higher (about 25 % consumption portion of most key world resources). Finally, some of those authors (there are many more) who have echoed a need for a more conserver urban society for America (and elsewhere), present and past, include: Alvin Toffler, John Ralston Saul, Noel Chomsky, Suzuki, Charles Reich, John K Galbraith, and Jane Jacobs. Can New York City or any urban area meet these writers challenges?

Universal thinking as noted above foresaw the potential benefits in thinking about the earth's resources and the need for conservation. This thinking led to many writings and community actions being taken up for conservation causes. To a degree a similar thinking in the 2000s, with roots over previous last 40 years, is represented in the green movement. This movement was in the early 1970s considered by most as restricted to a minority of people. By the 2000s it was evident that green policies needed to be considered in many areas (i.e. urban planning, energy consumption, greenhouse gas reductions, water planning and natural resource conservation). Within the urban planning area there continues moves to a more conserver (e.g. sustainable) society. The trend across most urban areas in the world over the last decade has been towards planning more sustainable communities. New York past Mayor Bloomberg, during his administration up to 2013, made many policy statements on the subject (www.nyc.gov).

3.3 Defining Sustainability

Given the writings on sustainability as outlined, throughout the United States and other countries, there has been a call for governments to respond to environmental issues. These issues include: need to renew older cities; impact of climate change; carbon emissions; pressures to accommodate new immigrant residents (note the recent debate on the ten million plus 'non-citizen' immigrants in the USA, a large sector residing in New York City, the city built on immigration as reviewed in Chap. 1); overpopulation of cities, the USA expected to continue growing faster than any western nation according to population projections by the United Nations; and, depleting resources (i.e. peak oil). Governments, in response, are trying to develop policies to address these concerns, often adopting strategies for sustainable communities (social/cultural, environmental, and economic). This chapter examines current urban planning schools, including land use planning; natural resource planning; and, sustainable urban planning. The chapter critiques the extent that current planning practices are moving towards a more coordinated and holistic framework in incorporating sustainability principles in our cities.

The subjects of sustainability, ecologically sustainable development (ESD) and ESD based urban planning have become of increasing interest worldwide, especially since the early 2000s. Concerns across countries have been aired on the failure of

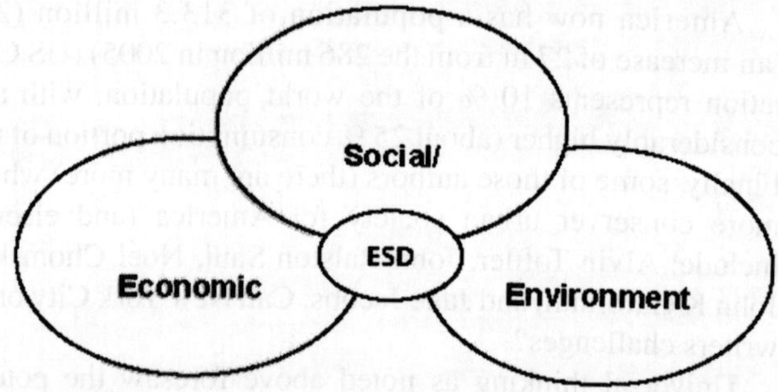

Fig. 3.1 ESD components – equal weight ((Source: UN Agenda 21) (1992a))

places to achieve environmental sustainability. These failures include for example: continued urban decline in many older cities (noted in Chap. 2 citing the demise of Bushwick in the 1970s and riots/fires of 1977); water shortages; rising global temperatures; unacceptable levels of greenhouse gas emissions; air and water pollution; increasing wastes (waste disposal solutions continue to plague New York City, having to truck its waste to locations outside the City); peak oil (50 % world's oil resources expended, thus challenging car dependent cities and need for alternatives, including more cycling and walking); and health virus pandemics (i.e. bird flu and swine virus). Given these failures and the resulting impacts on populations, the authors here focus on answers that could partly come from adopting environmental sustainability principles.

While worldwide expressions of concern about deteriorating environments continued in the 2000s, the writings on the subject accelerated. By 2008 convincing scientific evidence had pointed to the validity of concerns as raised. Writers who had been adding their voices and presenting evidence of environmental problems include: Lomberg (2004), Roberts (2004), Aplin (2006), Flannery (2006), Gore (2006, 2007), Stern (2006), Suzuki (2006), Grosvenor (2007), and Lovins (2012). The writers' comments often tied these highlighted environmental problems to the causal factor of uncontrolled urban growth. The thrust of this phenomenon in many countries is the deteriorating old industry cities and the outward expansion of cities into new growth areas. New York City (especially Brooklyn, including Bushwick) experienced this circumstance as this phenomenon as much as any industrial city as illustrated in Chap. 2.

Throughout the 1980s and into the first decade of the 2000s there have been numerous definitions and interpretations of the expressions 'sustainable', 'ESD' and 'sustainable urban planning'. 'Sustainable', as defined in the Macquarie Dictionary, is 'to provide the means of supporting life in a balanced way' (Macquarie Dictionary 2005). 'ESD' definition under the United Nations *Agenda 21* (Principle 3) (UN 1992a) is "development fulfilled equitably to meet developmental and environmental needs of present and future generations" (Fig. 3.1). The three ESD components (social/cultural, environmental and economic) are now widely accepted (UN 1992b). The social/cultural component can be taken as two layers for ease of analysis in any assessment of ESD. The components are also often interpreted as the Triple Bottom Line (TBL). The TBL (Benn et al. 2004) requires companies to balance the components of sustainability.

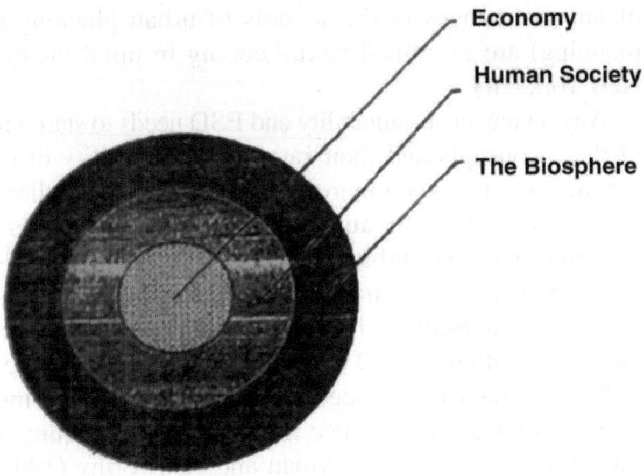

Fig. 3.2 ESD components – balance to environment (Source: Peet 2002)

There are, however, other models of sustainability (Aplin 2006). While the above model shows the three ESD components of equal importance (hence called the 'equal weight' model), there is an oft quoted version (Peet 2002) (Fig. 3.2). This version places the biosphere (environment) as more important over human society (social/cultural) and economic. Hence this model is referred to as the 'balance to environment' model.

In summary, the two models noted present a basis to start an analysis of sustainability and ESD. There are questions the planner faces in looking at these subjects within local areas, including; how to measure impacts in local areas; how to use indicators of sustainability measurements; how can indicators fit different sized areas; and, how sustainability reports can (with recommendations and monitoring component) be adopted.

The term 'local area' can be interpreted as an area that people can navigate by walking. A local area encompasses variable sized areas, including for example: precinct; institutional land (i.e. university); residential or industrial estate; seniors' area (i.e. retirement village); cultural land (i.e. immigrant groups); and neighborhood. A 'district' contains a defined number of 'local areas' and is usually navigated by car (rather than walking). New York City is a city built on a proliferation of local areas developed in the 1800s (called 'villages' or 'neighborhoods') as outlined in Chap. 1.

3.4 Urban Planning Background

Since post WW2, a great number of older urban settlements in the USA and worldwide have deteriorated, while others have expanded exponentially beyond metropolitan boundaries. One challenge, to engage this circumstance of growth, is for cities is to implement ESD based urban planning practices. This type of urban planning aims to incorporate ESD criteria across the full range of ESD components – social/cultural, environmental and economic. There are ever emerging trends and concepts of governments moving to adopt sustainability principles in urban planning. Changes in the schools of urban planning (land use and sustainability planning) are examined next, keeping in mind the circumstances of planning in New York City.

Any review of sustainability and ESD needs to start with examining the foundations of these concepts and their rationale. The ability of cities to meet ever changing social, economic and environmental needs is a challenge for governments. By the late 1990s there were authors calling for authorities to introduce sustainability principles into their urban planning. Trainer (1998) stated that society needs to find a more balanced and simpler way of building cities and neighborhoods. He states there are finite limits of resources available to build these areas. He thus argues for the incorporation of ESD criteria into the way cities are planned and renewed. Troy (1998) emphasizes the need for all levels of government, from local government to international, to introduce measurements of equity and efficiency in moving to sustainability of cities. Newman and Kenworthy (1999) introduce a sustainability framework around designing cities and neighborhoods that were not dependent on the automobile. Taking an overview, Saul (2002) argues that cities and the general functioning of societies within those cities have lost their equilibrium [equilibrium is the state of rest due to the action of forces that counteract each other (Macquarie Dictionary 2005)]. Lowe (2005) and Lovins (2012) advocate changes in the way our institutions (including urban planning) operate to heed the urgency to act on environmental matters affecting our cities and neighborhoods. Finally, Rauscher and Momtaz put a case to adopt a framework for planning sustainable communities (Rauscher and Momtaz 2013).

Other authors emphasize the importance of the social/cultural component of ESD. Stocker and Burke (2006), for example, places a high importance on place based sustainability education and the sharing of community wisdom. This emphasis on the social/cultural ESD component is further progressed by Hillier (2005), van de Kerkhof (2005) and Walsh and Mitchell (2002). Within the social/cultural ESD component questions of governance and the political processes also arise. Newman and Kenworthy (1999) explores this aspect around the subject of applying sustainability criteria for planning cities and neighborhoods, in particular transport. Other authors, including Kemp et al. (2005), Petschow (2005), and Voss (2006), have focused on specific aspects of governance incorporating an ESD approach. These authors emphasize the importance of flexibility in decision making and the ability of governments to take on new challenges raised by the public and or business sectors.

ESD related documents (i.e. protocols, acts and reports) represent the key advances of sustainability. These documents usually refer to sustainability criteria, consisting of: ESD principles; ESD goals; and, indicators of sustainability (Fig. 3.3). These sustainability criteria are contained within (and sometimes the bases of) documents at all levels of government (international, national, state and global) and non-government organizations.

There are a number of key words (e.g. principles, goals and indicators of sustainability) connected with ESD and contained in most documents on the subject (defined in the glossary). The adoption of ESD principles leads to goals and in turn leads to indicators of sustainability. 'ESD principles' were introduced at the

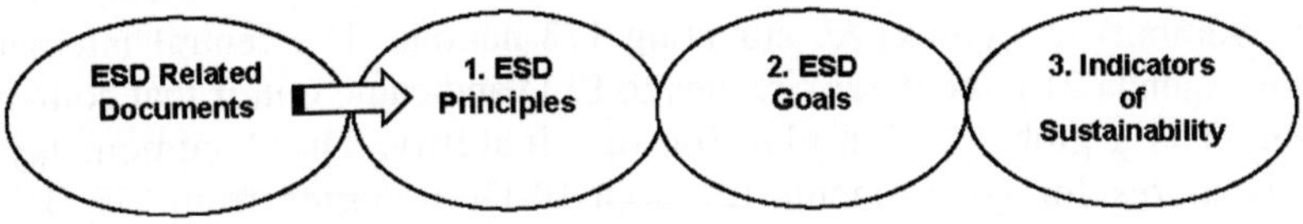

Fig. 3.3 ESD related documents and ESD criteria

international level at the *United Nations Conference on Environment and Development Report* (UN 1992b) (herein called the Rio Summit Report). The Rio Summit Report contains a declaration with twenty seven key principles varying from 'human beings are at the center of concerns for sustainable development' (Principle 1) to 'peace, development and environment protection are interdependent and indivisible' (Principle 25). ESD principles have been advanced within a number of other documents completed at different levels of governance. These levels include for example: (a) international (i.e. *Political Declaration and Plan of Implementation*) *(Johannesburg)* (UN 2002b); (b) national; and, (c) state.

Authorities, having adopted ESD principles, can then adopt ESD goals. Low (2000) argues that adopting ESD goals is crucial to confronting cities that are depleting resources without limit. He spells out how cities can adopt these goals and as a result ensure renewal of cities and preservation of resources for future generations. Equipped with ESD goals, authorities are in a position to adopt the key ESD components for measurement (called 'indicators of sustainability'). These indicators enable ESD goals to be measured and monitored at city or neighborhood levels.

At all levels of government, an increasing number of ESD related documents are being adopted. These documents include ESD related protocols, acts, agreements, reports and programs. These documents are produced at different government levels, including: international (UN); national; state; and, global non-government organizations (NGOs). The review now focuses on key documents that relate to ESD based urban planning, covering 41 years (1972–2013). Most of these documents have on their release been examined and at times adopted by city administrations, including New York City and sometimes at the City borough level, such as the Borough of Brooklyn.

3.5 International ESD Related Actions

International ESD related documents (particularly those applying to the natural and built environments), and affecting planning in cities such as New York, have increased in numbers and subject area. The first international ESD related document was the *Declaration of the UN Conference on the Human Environment* (UN 1972). This declaration was the first multi-national agreement to provide guidelines for nations to move to more sustainable human environments (social/cultural, environmental and economic). Twenty years after this declaration the *United Nations Conference on Environment and Development Report* (UN 1992b) (referred to as the Rio Summit) was convened, attracting 178 nations. The central international protocol, Agenda 21 (UN 1992a), relates to ESD and came out of that conference. Agenda 21 is a global action plan for sustainability. The document lays out key procedures for governments to adopt ESD strategies from city level to neighborhood level.

In moving beyond Agenda 21, the UN developed *Local Agenda 21* (UN 1993) from one chapter (Chapter 28) of Agenda 21. Local Agenda 21 is a mechanism to encourage greater involvement by local authorities in delivering Agenda 21 programs. In 1997 *Local Agenda 21 Model Communities Program* (UN 1997b) was adopted. This program aids local government in implementing the Local Agenda 21 program. The Model Communities Program documented those local government areas that had adopted Local Agenda 21 planning processes for sustainable development.

By the year 2000, the UN became aware that new directions were needed to assist local authorities in adopting ESD strategies. The UN thus created the *United Nations Commission of Sustainable Development* (UN 2000a) as a peak body to further the aims of ESD. To assist local government directly the UN adopted the *United Nations Sustainable Cities Program* (UN 2000b). This program provides guidelines to authorities on adopting ESD criteria (i.e. principles, goals and indicators of sustainability). The international ESD related document that continues to generate major public debate is the *Greenhouse Gas Reduction Agreement* (known as Kyoto Protocol) (UN 1997a). This protocol sets targets within a timeframe for greenhouse gas emission reduction for signatory countries. This protocol remains central to countries in cooperating on greenhouse gas emissions such as CO_2. The *Stern Report* (2006) updated the statistics on impacts of greenhouse gas emissions from an international perspective. The report contains recommendations to countries on amounts and time lines for CO_2 emission reductions. Nations have acted more swiftly following the release of this report, with many adopting CO_2 emission limits and payments or taxes for emission quantities. Most of these measures have been inclusive of renewable energy programs in older urban areas as well as outer city areas, including New York City.

Seeking a review of progress on Agenda 21 and subsequent protocols, the UN convened in 2002 the *World Summit on Sustainable Development* (UN 2002a). The World Summit adopted the *Political Declaration and Plan of Implementation* (UN 2002b) to provide further guidelines for local authorities to take appropriate actions towards ESD programs. This declaration addresses the need for authorities to give greater attention to areas of sustainability such as disaster management and climate change. In nations such as New Zealand plan making was changed to accommodate sustainability principles (Ericksen et al. 2004). In Canada, Montreal has exemplified initiatives in incorporating sustainability principles into urban planning. Brown (2006) assesses the extent that these principles are being incorporated into Montreal plans. New York City past Mayor Bloomberg instigated a number of sustainability guidelines into the City Council within the Dept of City Planning (2013). He concluded that these plans would lead to sustainable development outcomes in most instances. The recent initiative as noted in the following statement by the City (in this area joining forces with the State of Connecticut) reinforces the City's intentions in this area of sustainable urban planning (City of New York www.nyc.gov) (2013).

Start – taken from web (13 Nov 2013):

The Department of City Planning is participating in an unprecedented, bi-state collaboration of cities, counties and regional planning organizations who have united to form the New York-Connecticut Sustainable Communities Consortium. The Consortium was awarded $3.5 million in funding in the inaugural year of the U.S. Department of Housing and Urban Development's (HUD) Sustainable Communities Regional Planning Grant Program. The consortium will undertake a combination of regionally and locally oriented initiatives to promote sustainable, transit-oriented development throughout the region. Initiatives DCP leads under the Sustainable Communities program include three studies: a coordinated neighborhood planning study in East New York, Brooklyn; an evaluation of land use and transportation opportunities near Metro-North Stations in the Bronx; and several activities to advance citywide strategic planning efforts for building climate resilience.

End – taken from web (13 Nov 2013)

3.6 International Non-government Organizations and ESD

In addition to governments producing ESD related documents, many international non-government organizations (NGOs) have also produced documents to assist communities to move towards adopting sustainability planning approaches.

The NGO sector has been producing an ever expanding volume of ESD related reports since the early 2000s. These organizations to date have complimented the UN's efforts in ESD education, research and advocacy. The organizations include: International Institute for Sustainable Development; Sustainable Communities Network; Sustainable Cities; and, Urban Futures. This advocating is often primarily applicable to the interests and geographical areas that these organizations are active within.

The first NGO to produce an ESD related report was the *Club of Rome* (1972). This report by an eminent group of experts expressed environmental concerns that have only partly been taken up by governments since 1972. An overview of how these concerns were addressed over 40 years ago was taken up by Meadows et al. (2005). She argues that during this time few of the concerns were addressed. She goes further to outline the means to a transition to a sustainable system and notes the tools for that transition. International environmental NGO organizations adopted the *World Conservation Strategy* (IUCNNR 1980). These same groups put together a report containing an expanded set of principles for ESD entitled *Caring for the Earth – a Strategy for Sustainable Living* (IUCNNR, UNEP, WWF 1992). ESD principles in this report include: respect and care for the community of life; improve the quality of human life; and diversity; minimize the depletion of non-renewable resources; change personal attitudes and practices; and, enable communities to care for their own environments. The European Commission in 2001 launched a campaign for a sustainable European network of cities and towns. The aim was to group more than 540 local authorities in formulating policies to promote sustainability.

Environmental groups have introduced new environmental terms in promoting ESD principles, such as ecological integrity [ecological integrity means all natural processes and interactions within an ecosystem are maintained] (Bell 1994). Community interest ranged up to a dictionary being published on sustainability (Aplin 2006). A book defining these ESD principles that the NGO sector have called for in the past are summarized by Beder (1996). These have included: social equity; qualitative development; pricing environmental values; natural capital and sustainable income; and, wider community participation within ESD policy making.

Observations can be drawn from assessing the timeline of accumulated years for NGO sector actions on ESD related reports. Generally, the NGO sector is increasingly pro-active in alerting governments. Key NGO documents have contributed to the debate around national actions on climate change (Stern 2006). Time delay of years and sometimes decades exist, however, in governments adopting NGO recommendations. At the same time, the NGO sector is increasingly producing detailed reports reflecting the need for government actions. The take up by the NGO sector of Al Gore's *An Inconvenient Truth* (2007) is a measure of this sector's work in educating the public on environmental issues. The worldwide education of groups of 'climate change educators' trained under Al Gore is testimony to the take up of sustainability issues by the NGO sector.

It was the climate change concerns, however, that continued to drive public interest from 2003 to present (2013). Several states in the USA were also adopting their own greenhouse gas protocols. In addition the mayors of a number of USA cities, including New York City, agreed to cooperate on CO_2 emissions.

3.7 ESD Based Planning Frameworks

Having examined the basis of sustainability and ESD (e.g. the environmental aspects), the focus now shifts to examining ESD based urban planning frameworks, especially schools of urban planning thought. Within cities urban planning consists broadly of 'land use planning' and 'natural resource planning'. Increasingly authorities are also moving towards 'sustainability planning', where the application of ESD criteria is essential. Planning in the City of New York has experienced all of these transitions, often taking the lead and at other times being led by community advocacy groups (especially within the urban renewal areas) (Angotti 1999).

3.7.1 Land Use Planning

Land use planning incorporates the application of land development and conservation principles to resolve the use of land. This urban planning school of thought encompasses, for example: the type of physical layout desired; accommodating, environmental, social/cultural and economic; and, ensuring services at different levels such as localities, districts, cities and regions. Urban planning land use schools of thought date back to the Roman Empire, however the book focuses on current and recent historical changes in land use planning schools.

Other schools of land use planning thought emerged in the 1990s. Bertugia (1994) argues that we can model the city based on the performance we expect from the city. Rose (1997) continues with that argument and states that authorities need to look at land use planning not in terms of zonings but new dimensions of creating healthy communities, locally and globally. These principles also reflect Archibugi's (1997) argument that planners need to know the likely impacts of urban growth. He suggests the means to move to more planned (and thus more sustainable) communities. Likewise, Forster (1999) pushes for reforming land use planning to achieve more sustainable cities. The examples of several Brooklyn major planning initiatives over the last 10 years (to 2013) will be covered in latter chapters.

Key urban planning based schools of thought influencing the take up by authorities of ESD criteria include: *compact cities* and *new urbanism*; *smart growth; and, eco-city planning and eco-villages*. These recent land use schools of thought are briefly commented on below, including comment on the New York City position on these schools.

Compact cities, as the name suggests, aims for higher density development around transport nodes. The document aimed to slow urban sprawl by increasing the allowable urban densities. This was a reversal of past strategies (after the Second World War) that encouraged urban sprawl. It is under compact cities that ESD principles gained attention in cities such as New York, including Brooklyn (incorporating Bushwick). New York of course represents one of the most compact places in the western world.

New urbanism emphasizes building neighborhoods with a diversity of residential, commercial and light industrial land uses in close proximity. This urban planning school has been popular throughout the 1980s to current times. New urbanism seeks to "reform urban design processes, restore life in urban centers and rely less on motor vehicles" (Crofts 1998, p. 28). The American planners argue new urbanism through institutions such as the American Planning Institute (1999). In Australia, Newcastle City Council responded by applying concepts of new urbanism incorporating sustainability principles (McKay and Rauscher 2007). The authors outline the progress and setbacks of the Newcastle Council over several years starting with the Council's attendance at the Rio Summit in 1992. In New York City the Planners Network has promoted new urbanism principles for several decades, taking up many issues with the city administrators. The next two chapters explore many Brooklyn, including Bushwick, planning projects initiated by the City of New York, not for profit community groups, and developers.

The notion of *smart growth* followed new urbanism as a complementary school of planning thought in the early 1990s in the United States (Urban Land Institute 1995). This school aims to limit urban sprawl through improved land use and transport policies. The smart growth movement emphasizes greater efficiencies of urbanization through the incorporation of a wide range of ESD based urban planning principles (i.e. energy, water, and transport). Newman (1998) argues that sustainable transport will be the most important ESD principle for local government and states to accommodate within their urban strategies. Stillwell (2000) also challenges authorities in drawing comparisons between American sprawl compared with Australian. He offers policy directions leading to more efficient urban development across Australia to combat sprawl. Older inner city areas such as those in Brooklyn thus become increasingly important as a city aims to implement smart growth.

A further land use urban planning school of thought is *eco-city* planning. Walker (1997) argues that a range of tools are being applied within the concept of eco-cities, providing authorities adopt the appropriate strategies. Hollick (1998) critiques the ESD lessons learned by eco-villages that have been functioning over several years and argues their attributes. Engwicht (1999) argues that you can create eco-communities by adopting ESD criteria at the street design level. Barton (2000a) goes beyond streets and argues that there is potential for whole neighborhoods to become sustainable communities (environmental, social/cultural, and economic components). In addition, Rauscher and Momtaz (2004) outline tools for ESD based urban planning that practitioners could utilize. Likewise, an outline of the setting up of a number of eco-villages in Australia are provided by Rauscher and Momtaz (2013) in a recent book *Sustainable Communities: A Framework for Planning*. The most recent eco-village in Australia currently being established is the Narara Eco Village (NED), within an hour north of Sydney on the Central Coast. Cities and localities within cities such as Bushwick can adapt a range of planning parameters from the experiences of eco-villages. A number of these villages are established outside the New York City metropolitan area.

3.7.2 Natural Resource Planning

While urban planning incorporates land use and natural resource planning it has been natural resource planning that has had a major influence on urban planning from the 1970s to date. Recent natural resource planning schools of thought influencing urban planning as summarized below include: eco-accounting; eco-design; ecological foot printing; limits to growth; and, measurements of progress. The City of New York has in various recent policies tried to incorporate many of these schools, as a search on the City's web will illustrate (City of New York www.nyc.gov) (2013).

Eco-accounting complements the bioregional planning and consists of measuring the impact of development on a 'loss and gain balance sheet' basis. The Australian Conservation Foundation (ACF) (1990), for example, utilizes eco-accounting in its argument that taxes are a true value for measuring the cost of paying for development impact. The work of Henry George in the late 1800s (as introduced in this book's acknowledgments) is very relevant to eco-accounting. Georgist Societies today continue to work with government administrations such as the City of New York to illustrate the means of just collection and distribution of taxes (Drake 2010).

These eco-accounting principles, Georgists argue, ensure adequate revenue is raised to meet the needs of an urban population, including areas that need renewal such as Bushwick.

Eco-design goes beyond eco-accounting by applying ecological principles to buildings and whole neighborhoods. Barton (2000b) argues that planners need to measure and apply principles of eco-design in achieving sustainable settlements. Inoguchi (1999) points to the prospects of eco-design in all aspects of society. He argues that in time we can create 'eco-societies'. The work of the Rocky Mountain Institute (1998) illustrates developments that can be planned on a green design basis. The Rocky Mountain Institute demonstrates its philosophy within demonstration projects in Colorado, USA, and around the world. A number of new low and high rise buildings, commercial and residential, built in New York City over the last several decades have been innovative in adapting eco-design principles (Department of City Planning, City of New York, http://www.nyc.gov/html/dcp). The background to eco-design as practiced in New York City was provided in Chap. 2, giving examples in Brooklyn (including Bushwick) and Manhattan.

Ecological foot printing is another tool for natural resource planning. Foot printing provides a means of measuring the impact of human activities on an individual and on a cumulative society basis. Packard (1991) showed in the early 1990s that authorities could plan settlements naturally and thus avoid footprint impacts. The UN Office of Economic Cooperation and Development (UN 1995) produced a handbook to help authorities around the world to utilize renewable energy approaches to reduce footprints. One of the biggest city footprints in the world is New York City (extending into several other States), in spite of urban efficiencies that come from higher residential densities.

A limits to growth suggests there are a finite number of people that is tolerable in an environment. The introduction of people beyond a limit places undue stress on inhabitants and that environment. The Australian Theosophy Society as early as 1914 was examining the environmental impact that communities had around the world (Theosophy Society 1914). The Club of Rome (1972) originally put forth the concept of limits to growth. From another perspective, Fincher (1998) argues that communities need to reframe the questions we ask to assess population capacities. Debate over peak oil and peak coal encompasses arguments of limits to growth (Lomberg 2004). Finally, Swamy (2001) takes a broader view and argues that growth needs to be looked at against the spiritual values we place on the environment and cities. New York City, given its impact reaching Long Island and other States, continues to debate the questions of the impact of the city's population what limits to growth could apply. That growth for example extends to Long Island (attached geographically to New York City as shown in Chap. 1) and having a population over four million (within counties of Nassau and Suffolk).

Complementing the 'limits to growth' school of thought is the concept of *measurement of progress*. Measurement of progress introduces a means of adopting criteria that can be used to gain an overall rating of progress (i.e. achieving healthy and sustainable communities). As early as 1985 Tisdell (1985) argued that there are conflicting views on what constitutes progress among economists and ecologists.

He argues that an approach to sustainable development would assist in defining the differences in these views. This question was later examined by Redefining Progress Inc. (2000). The continued aspect of poverty co-existing with prosperity in places like New York City (including Brooklyn and Bushwick) challenges the City Council and the State of New York. The work of Henry George as noted earlier and in book's dedication, in his classic *Our Land and Land Policy* (George 1871) is applicable to this question today. The work by George was updated in 2010 by the New York based Robert Schalkenbach Foundation (Drake 2010), addressing current circumstances progress and poverty in New York City and elsewhere in America.

3.7.3 Sustainable Urban Planning

There has been emerging since the 1990s a sustainable urban planning school of thought. This school of thought partly relies upon frameworks being developed to guide ESD based urban planning. Grant et al. (1996) proposed a planning framework for the protection of landscapes and ecosystems for residential environments. Given there have been a number of eco-villages already established, research was conducted to review how these villages incorporated sustainability principles (see also earlier comments in this chapter on eco-cites). In general, the eco-villages subscribe to Grant's et al. (1996) frameworks. Finally, examining land requirements, Beatley and Manning (1997) argues for land to only be consumed on a sparingly basis. He introduces 'sustainable places' considering not just physical layout of an area but the way the community operates. Some of the planning initiatives in places like Downtown Brooklyn and Southeast Brooklyn (including Coney Island), and whether those initiative do (or do not) address sustainability principles is canvassed in Chap. 5.

Other sustainable urban planning frameworks were outlined in the late 1990s. Zackary (1999) argues that indicators of sustainability need to provide the required guideposts (contained in measurements) for a local planning framework. He states that too often urban development decisions are made without testing likely impacts (i.e. infrastructure planning affecting whole neighborhoods). Likewise, Stimson (1999) argues for a whole of government agreement of adopting an urban planning framework of indicators of sustainability. Stimson applies his framework to land use and natural resource planning. He demonstrates how indicators can become part of a government's decision making. Ravetz (1999) adds to the Stimson work by proposing a framework of integrated strategic management methods and tools, including sustainability indicators, for neighborhoods, cities, and regions. Finally, Crilly and Mannis (2000) introduces a framework of sustainable strategic management systems as a means of sustainable urban planning.

As noted above by the late 1990s, sustainability frameworks for urban planning were being outlined. Fremantle (Newman 1998), Newcastle (McKay and Rauscher 2007), Sydney (Rauscher and Momtaz 2013) were Australian cities adopting sustainability approaches to planning. Beyond Australia, Manchester City Council, England, became a leader amid local government in selecting ESD indicators of sustainability. The City of Manchester (1999) adopted a range of indicators of sustainability within urban planning policies across the city. The city council established a matrix system of indicators to measure how ESD components (social, environmental and economic) were being met. Manchester added the concept of 'needs verses wants' in applying the indicators. This enabled the authorities to make decisions based on a priority of needs to protect the environment. Cities throughout the world are now conferring through various member groups of city administrations and often sharing the means of introducing more sustainability based urban planning (i.e. in transport planning, including cycleways; and, green corridor planning). Appendix 1 lists a number of web sites of New York City based organizations contributing to furthering these aims.

Throughout the early 2000s authors continued to develop frameworks that incorporated sustainability principles into urban planning. Crowe (2000) outlined this movement in arguing that it aims for a more civil society through land use and natural resource planning being based on sustainability principles. Crilly and Mannis (2000) developed a framework for spatial urban design indicators and a methodological tool kit. Crilly and Mannis (2000) calls upon authorities to think holistically about complex urban systems and suggests a locality specific approach to explain the totality of the system. Within Ireland, O'Regan et al. (2002) reported on the Irish Environmental Protection Agency developing a framework that relied upon spatial policies in developing sustainable regions. The aim here was to create optimum sized settlements that least harmed the environment.

Moving from England to the USA, the Urban Land Institute (2000) produced a framework on sustainable urban planning for American planning application. Works, such as the latter, have assisted the City of New York Council. In addition, the *Field guide to the Natural World of New York City* (Leslie Day 2007) indicated that New York City could be the most biologically diverse city in temperate America. Leslie draws attention to the fact that the five boroughs sit: a. directly under the Atlantic migratory flyway; b. at the mouth of a 300 mile long river; and, c. on three islands – Manhattan, Staten Island and Long Island. She concludes that New York City can be as interesting environmentally and as exotic as that of any place on earth. She comments on (and lists) New York City organizations that work with urban planning authorities to protect the natural environments of the City in moving to a more sustainable city.

Moving from the Irish experiment to the UK, Phillips (2003) developed a framework of assessment for determining the energy and environmental capabilities of a local area for sustainable development. Phillips puts forth the framework to avoid environmental consequences of ill-considered development. Finally, Spencer (2005) highlights the Scandinavian 'eco-municipalities'. Spencer describes the Stockholm's 'The Natural Step' framework as places that have voluntarily committed to integrating sustainability principles to create green solutions. The initiatives by New York City in bicycle path planning throughout the five boroughs have been a reflection of the City's response to community advocacy for changes. They launched a long-term sustainability plan called *PlaNYC: A Greener Greater New York*

(City of New York 2008) Other measures are being taken by the City in the area of sustainable transport planning as green alternatives to the dependence by many on the automobile.

By the mid-2000s other frameworks for sustainable planning focused on the built environment. Sahely et al. (2005) develops a framework for urban infrastructure systems based on feedback mechanisms (using indicators of sustainability) between that infrastructure and the surrounding environment. Brandon (2005) argues that authorities everywhere need to be able to evaluate the built environment for the level of sustainability they wish to achieve. He sets a context for evaluating sustainable development through frameworks, including better assessment methods and management systems. His approach is to set standards within models covering a range of built environment outcomes. Hyde et al. (2007) considers the use of an environmental brief to gain more sustainable built environments. He argues for design strategies to create environmentally sensitive buildings. Finally, in New York City green buildings are becoming an increasingly common element in the City, according to the City's *The Green Building Handbook* (City of New York 2010).

The international debate on capping greenhouse emissions (Aplin 2006) is an example of a sustainability issue that will influence land use and natural resource decision making. Measures taken under greenhouse emissions capping are likely to impact on industry locations and urban renewal (land use planning) and extent of allowable vegetation removals (natural resource planning). The carbon credit systems being proposed at national, state and private industry levels is likely to affect urban growth decision making (Gore 2007). The issue is summarized by Newman and Jennings (2008) in discussing cities as ecosystems and bioregion environments. Newman here argues we need to take cues from the living systems for sustainability strategies in fostering a sense of place. There has been considerable take up of this call by New York City community and educational bodies, though the applications can be trying as discovered in the coming chapters.

3.8 Chapter Summary

This chapter summarized the state of the world's environments and the need to plan using sustainability criteria in all cities, with a focus on New York. The chapter reviewed terms of sustainability and its components. The chapter also summarized sustainability in the context of urban planning. There is a wide divergence of definitions of sustainability and ESD. It is clear, however, that there is increasing awareness among city administrations to try to reach a consensus on sustainability definitions.

Trends of governments moving towards ESD based urban planning frameworks are evident. This movement is detected within all three planning schools of thought (e.g. land use planning, natural resource planning, and, sustainable urban planning). In summary, governments and planning schools are moving towards incorporating sustainability principles in their planning (as in many of the examples cited in this chapter). There remains, however, a need for a comprehensive framework to adopt sustainability planning principles. New York City's urban renewal planning is examined in this context in the next chapter. The chapter looks at the creation and operation of New York City community boards and an array of planning instruments, especially the 197-a Plan process.

References

American Planning Institute (1999) New American urbanism – reforming the suburban metropolis. Cambridge University Press, New York

Angotti (1999) New village journal. http://www.newvillage.net/Journal/Issue1/1angotti.html

Aplin G (2006) Global environment in crises: an Australian perspective. Oxford University Press, Sydney

Archibugi F (1997) Ecological city and city effect: urban planning requirements for the sustainable city. Urban Policy Res 17(4):340–342

Australian Conservation Foundation (1990) Environmental tax reform. Author, Melbourne

Barton H (2000a) Sustainable settlements – a guide for planners. University of West England, Bristoe

Barton H (ed) (2000b) Sustainable communities: the potential for eco-neighborhoods. Earthscan, London

Beatley T, Manning K (1997) The ecology of place: planning for environment, economy and community. Island Press, Washington, DC

Beder S (1996) The nature of sustainable development. Scribe, Newham

Bell F (1994) A-Z of ESD. Earth Foundation, Sydney

Benn S, Dunphy D, Griffiths A (2004) Corporate change for sustainability: the way ahead. Australian Conservation Foundation, Carlton

Bertugia CS (1994) Modeling the city: performance, policy and planning. Routledge, New York

Brandon PS (2005) Evaluating sustainable development in the built environment. Blackwell Science, Oxford

Brown D (2006) Back to basics: the influence of sustainable development on urban planning with special reference to Montreal. Can J Urban 1:18–23

City of New York (2008) PlaNYC: a greener greater New York. City of New York, New York

City of New York (2010) The green building handbook. City of New York, New York

City of New York (2013) www.nyc.gov

Club of Rome (1972) Limits to growth. Universe Books, New York

Crilly M, Mannis A (2000) Sustainable urban management systems. In: Williams K (ed) Sustainable urban form. EFN, London

Crofts D (1998) Place management in Newcastle City. Aust Plann 35(1):15–20

Crowe N (2000) Building cities – towards a civil society and sustainable environment. Art Media Press, New York

Day L (2007) Field guide to the natural world of New York City. The John Hopkins University Press, Baltimore

Department of City Planning (2013) City of York. City web site under Department of City Planning: http://www.nyc.gov/html/dcp

Drake B (2010) Henry George progress and poverty. Robert Schalkenbach Foundation, New York

Engwicht D (1999) Street reclaiming: creating livable streets and vibrant communities. Pluto Press, Sydney

Ericksen N et al (2004) Plan-making for sustainability: the New Zealand experience. Ashgate, Aldershot

Fincher R (1998) Population questions for Australian cities: reframing our narratives. Aust Geogr 29(1):18–23
Flannery T (2006) Weathermakers. Collins, Sydney
Forster C (1999) Australian cities, 2nd edn. Oxford University Press, Sydney
George H (1871) Our land and land policy. Henry George Society, New York City
Gore A (2006) Earth in balance: ecology and human spirit. Nodale 2006, New York
Gore A (2007) An inconvenient truth. Wiley, New York
Grant J, Manuel P, Joudrey D (1996) A framework for planning sustainable residential landscapes. J Am Plann Assoc 62(3):331–345
Grosvenor M (2007) Sustainable living for dummies. Wiley, Sydney
Hillier J (2005) Mind the gap. In: Hillier J, Rooksby E (eds) Habitus: sense of place. Ashgate, Aldershot
Hollick M (1998) Sustainable communities: lessons for aspiring eco-villages. Praxis, Sydney
Hyde R et al (2007) The environmental brief: pathways for green design. Taylor & Francis, Abingdon
Inoguchi T (ed) (1999) Cities and the environment: new approaches for eco-societies. United Nations University Press, New York
International Union for Conservation of Nature and Natural Resources (IUCNNR) (1980) World conservation strategy. IUCNNR, Switzerland
IUCNNR, UNEP, WWF (1992) Caring for the earth: a strategy for sustainable living. IUCNNR, UNEP, WWF, Gland
Kemp R, Parto S, Gibson RB (2005) Governance for sustainable development: moving from theory to practice. Int J Sustain Dev 8(1–2):12–30
Lomberg B (2004) Global crises, global solutions. Cambridge University Press, Cambridge
Lovins LH (2012) The way out. Hill and Wang, New York
Low N (ed) (2000) Consuming cities: the urban environment in the global economy after the Rio Declaration. MIT Press, Cambridge
Lowe I (2005) A big fix: radical solutions for Australia's environmental crisis. Black Inc., Melbourne
Macquarie Dictionary (2005)
Manchester City Council (1999) The state of the city region. Manchester Press, London
McKay D, Rauscher R (2007) Devolving sustainability policy to local government: a case study of the City of Newcastle. J Aust Pol Econ 58:74–94
Meadows D, Randers J, Meadows D (2005) Limits to growth: the 30-year update. Earthscan, London
Newman P (1998) From symbolic gesture to main stream: next steps in local sustainability. Murdoch University, Perth
Newman P, Jennings I (2008) Cities as sustainable ecosystems: principles and practices. Island Press, Washington, DC
Newman P, Kenworthy J (1999) Sustainability and cities: changing automobile dependence. Island Press, Washington, DC
O'Regan B, Moles R, Kelly R, Raveltz J, McEvoy D (2002) Developing indicators for the estimation of sustainable settlement size in Ireland. Environ Manage Health 13(5):450–465
Packard S (1991) Planning settlements natural. Routledge Press, London
Peet (2002) Sustainable development: why is it so difficult? Pac Ecol 4(Summer 2003): 16–20
Petschow U, Rosenau J, Ulrich von Weizsacker E (2005) Governance and sustainability: new challenges for states, companies and civil society. Greenleaf Publishing, Sheffield
Phillips C (2003) Sustainable place: a place of sustainable development. Wiley, Chichester
Rauscher R, Momtaz (2004) Sustainable Area Planning (SAP) framework for sustainable transport in growth centers. In: 28th Australasian Transport Research Forum. Australasian Transport Research Forum 28, Sydney
Rauscher RC, Momtaz M (2013) Sustainable communities: a framework for planning. Springer, Amsterdam
Ravetz J (1999) City-region 2020: integrating planning for a sustainable environment. Earthscan, London

Redefining Progress Inc. (2000) Community indicators handbook. Cambridge University Press, New York
Roberts P (2004) The end of oil. Bloombury, Staffordshire
Rocky Mountain Institute (1998) Green development. Wiley, New York
Rose M (1997) Eco-city dimensions: healthy communities and healthy planet. New Society Publications, Gabriola Island
Sahely HR, Kennedy CA, Adams BJ (2005) Developing sustainability criteria for urban infrastructure systems. Can J Civil Eng 32(1):72–86
Saul JR (2002) On equilibrium. Penguin, New York
Spencer M (2005) Building sustainable cities. E Environ Mag 16(5):20–23
Stern R (2006) International climate change final report. Her Majesty's Treasury, London
Stillwell F (2000) Reshaping Australia: urban problems and policies. Pluto Press, Sydney
Stimson R (1999) Urban metabolism as a framework for sustainable development in SE Queensland metro region. Urban quality of life, Singapore conference proceedings, University of Singapore, Singapore
Stocker L, Burke G (2006) Chapter 25: Overlay mapping – a methodology for place based sustainability education. In: F Wooltorton, D Marinova (eds) Sharing wisdom for our future: environmental education. Self Published, Sydney
Suzuki D (2006) Suzuki David the autobiography. Allen Unwin, Vancouver
Swamy CRN (2001) A spiritual ecology. The Theosophist, August 2001, New Delhi
Theosophy Society (1914) Ideal communities. Herald of the Star, May 1914, Sydney
Tisdell CA (1985) Sustainable development – conflicting approaches of ecologists and economists. University of Newcastle Economics Department, Newcastle
Trainer T (1998) The simpler way. University of NSW Press, Sydney
Troy P (1998) Equity, environment efficiency: urban Australia. Urban Futures, May 1998
United Nations (1972) Declaration of the UN conference on the human environment. UN, New York
United Nations (1992a) Agenda 21. UN, New York
United Nations (1992b) United Nations conference on environment and development report (Rio Summit). UN, New York
United Nations (1993) Local Agenda 21. UN, New York
United Nations (1995) Urban energy handbook: good local practice. OECD/UN, Washington, DC
United Nations (1997a) Greenhouse gases reduction agreement report (Kyoto Protocol). UN, New York
United Nations (1997b) Local Agenda 21 model communities program. UN, New York
United Nations (2000a) United Nations Commission of Sustainable Development (CSD). UN, New York
United Nations (2000b) United Nations sustainable cities programme. UN, New York
United Nations (2002a) World Summit on sustainable development report. UN, New York
United Nations (2002b) Political Declaration and Plan of Implementation (PDPI). UN, New York
Urban Land Institute (1995) Smart growth. Cambridge Press, New York
Urban Land Institute (2000) Development practice of sustainable cities. Cambridge Press, New York
Van de Kerkhof M (2005) Learning and stakeholder participation in transition processes towards sustainability: methodological considerations. Technol Forecast Soc Change 72:733–747
Voss JP, Bauknecht D (2006) Reflexive governance for sustainable development. Edward Elgar, Cheltenham
Walker B (1997) Sustainable cities: concepts and strategies for eco-city development. EcoHome Publ., Los Angeles
Walsh F, Mitchell P (2002) Participatory planning. Planning for Country. Jukurrpa Books. Central Land Council
Zachary J (1999) Sustainable community indicators – guideposts for local planning. Wiley, New York

Chapter 4
Planning in New York City: Community Boards and Planning Instruments

4.1 Introduction

By 2013 Bushwick, like the rest of Brooklyn and New York City, was continuing to renew. One of the stimulants for renewal of neighborhoods throughout the City was the Community Planning Program of the City of New York as introduced in 1975. This included legislation for Community Boards and 197-a Plans. This chapter looks at this legislation and the review of its effectiveness in light of the movement to sustainable urban planning (subject of Chap. 3). In particular, the New York City Charter authorizing Community Boards to sponsor plans such as those under 197-a is examined. The revitalization of neighborhoods under these Boards and application of 197-a Plans is to be critiqued in Chap. 5.

4.2 Community Boards

Engaging the community in urban planning had its beginnings in the 1960s. Paul Davidoff, founder of Hunter College's urban planning program, wrote significant books on the idea of advocacy in urban planning. Perhaps aware of this community interest, as early as 1963 the New York City wrote into the City Charter the establishment of Community Boards. The city was divided into community districts with each governed by an advisory planning board. The boards consist of community residents appointed by the borough presidents. Plate 4.1 denotes the fifty nine (59) Community Boards covering the five boroughs of the city. The City set up a web site enabling residents and businesses to find the Community Board they come under (http://www.nyc.gov/html/cau/html/cb/cb.shtml).

Community Boards were first thought of in 1951 under Mayor Robert F Wagner. At the time these were called Community Planning Councils and set up in Manhattan. The councils played advisory roles on planning and financial questions. In time, and still under Wagner, Community Planning Boards were established in all boroughs

R.C. Rauscher and S. Momtaz, *Brooklyn's Bushwick - Urban Renewal in New York, USA: Community, Planning and Sustainable Environments*, DOI 10.1007/978-3-319-05762-0_4,

Plate 4.1 Community Boards of New York City (Source: Dept. of City Planning, City of New York 2013a, b)

under the 1963 New York City Charter. It was under Mayor John Lindsay (1970s) that the City went further and created Little City Halls in several districts. These 'little city halls' had district managers to supervise city services and 'service cabinets' of different city agencies representatives for local inter-agency coordination.

In 1975, it took the New York City voters to approve a new City Charter combining earlier programs to create the Community Board system. The Charter called for each Board to be allotted a district service cabinet and district manager to be appointed by the Board. Additionally, the 1975 Charter introduced the Uniform Land Use Review Process (ULURP). This required the Community Boards to review all land use applications. These included: zoning actions; special permits; acquisition and disposition of city property; and urban renewal plans.

Given the importance of Community Boards to planning at the local level in New York City (including eventually 197-a Plans) a summary of how the boards are structured and operate follows (refer to City of New York web under Community Boards http://www.nyc.gov/html/cau/html/cb/about.shtml#govt).

The City's 59 community boards are numbered independently within each borough. Community districts are defined by the Department of City Planning and

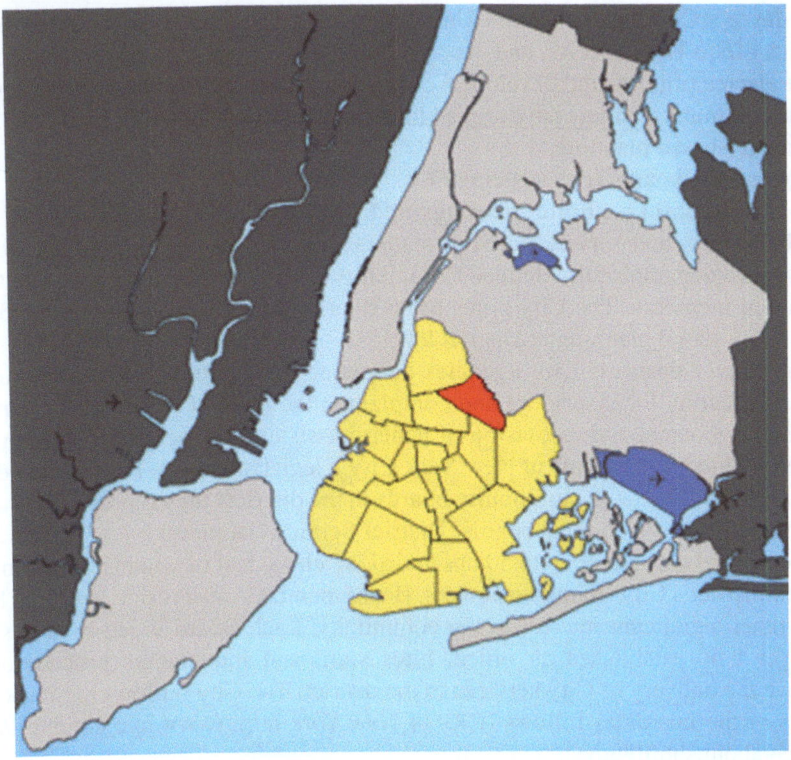

Plate 4.2 Bushwick in context of Brooklyn, Five Boroughs, and Long Island. *Left* to *Right*: with Bushwick center (*red*) (Clockwise) NYC Boroughs (*light gray*): Staten Island (south); Manhattan (west); Bronx (north of Manhattan); Queens (adjacent to Bushwick, east); and, Brooklyn (*yellow*); and Long Island (*dark gray* far east) (Source: Dept. of City Planning, City of New York 2013a, b)

are drawn roughly along the lines of one or more "neighborhoods," though these are more subjectively defined and may also span more than one community district. Plate 4.2 illustrates the location of the community board responsible for Bushwick. Left to Right: is Bushwick center (red) and clockwise are NYC Boroughs (light gray): Staten Island (south); Manhattan (west); Bronx (north of Manhattan); Queens (adjacent to Bushwick, east); and, Brooklyn (yellow); and Long Island (dark gray far east).

Community board members are selected by the Borough President according to the charter members need to be "among active, involved people of each community, with an effort made to assure that every neighborhood is represented," and must "reside, work, or have some other significant interest in the community," Half of the board members must be nominated by the City Council members representing that district, but are ultimately selected by the Borough President.

The District Managers of Boards play a crucial role in community board administration: maintaining the office, hiring staff, and surveying city services delivery.

The Mayor's Community Affairs Unit states that "The main responsibility of the District office is to receive and resolve complaints from community residents." While elected officials and developers often look to community boards for the voice of "the community," many have argued that the system does not effectively facilitate community-based planning.

Community boards are a system of local representative bodies in New York City. They hold monthly, public meetings and advise other city agencies on land use, budgetary, and service delivery matters. Each of the 59 community boards represents a geographically defined Community District and is made up of up to 50 unsalaried members. The City notes these Boards were an early attempt to foster community-based planning at District level.

The Mayor ensures: City agencies cooperate with community boards in all matters affecting local services and complaints; the level of financial support for community boards; and, provision of general assistance as needed. City Council elected members (51 in all in New York City's legislative body) are supposed to be closely involved with community boards in the districts they represent. Council members also serve on their boards' District Service Cabinets.

Each Board consists of up to 50 unsalaried members, half of whom are nominated by their district's City Council members. Board members must reside, work, or have some other significant interest in the community. Each Board is led by a District Manager who establishes an office, hires staff, and implements procedures to improve the delivery of City services to the district. The City outlines the responsibilities of the boards as follows (City of New York http://www.nyc.gov/html/cau/html/cb/about.shtml#govt):

Start: taken from web (13 Nov 2013):

Community boards have a variety of responsibilities, including but not limited to:

1. Dealing with land use and zoning issues.

   Boards have an important advisory role and must be consulted on the placement of most municipal facilities in the community. Applications for a change in or variance from the zoning resolution must come before the board for review, and the board's position is considered in the final determination.

2. Assessing the needs of their own neighborhoods.

   Boards assess the needs of their community members and meet with City agencies to make recommendations in the City's budget process.

3. Addressing other community concerns.

   Any issue that affects part or all of a community, from a traffic problem to deteriorating housing, is a proper concern of community boards.

End: taken from web (13 Nov 2013)

Finally, the City produced an introduction to Community Boards within a Handbook (City of New York 2013b) (www.home.nyc.gov). The Handbook focuses on their responsibilities of Boards within the larger framework of City government. The Handbook also outlines the responsibilities of the District Manager and District Service Cabinet. The Handbook's appendices include relevant sections of the City Charter and additional useful information for board members.

4.3 197-a Planning

It was in 1975 that the City Charter initiated local planning under Section 197-a. This tool gave community boards and other community associations the right to take an active role in planning. This move by the City gave community boards the opportunity to develop neighborhood plans. It was in 1989 that the City adopted rules establishing minimum standards for content of 197-a plans as well as a schedule for review.

The City of New York issued the *197- a Plan Technical Guide* (NYC 1997) (here in called the '197-a Guide') under Mayor Rudolph W. Giuliani in 1997. A handful of community boards had already successfully negotiated the planning processes under the 197-a Charter (1963). The Office of the Director of City Planning (Joseph B. Rose) in the Guide's covering letter reinforced the success of a 197-a plan as it could "build consensus within a community about its future direction, challenge conventional wisdom, and set the stage for beneficial shifts in city policy". Formal steps in the review and adoption of 197-a plans process are summarized in Plate 4.3.

A summary of the progress of communities utilizing the 197-a as noted by the City Administration, including adopted plans, is presented in Plate 4.4. There are 15 plans listed here in chronological order and indicating by column: a. Name of

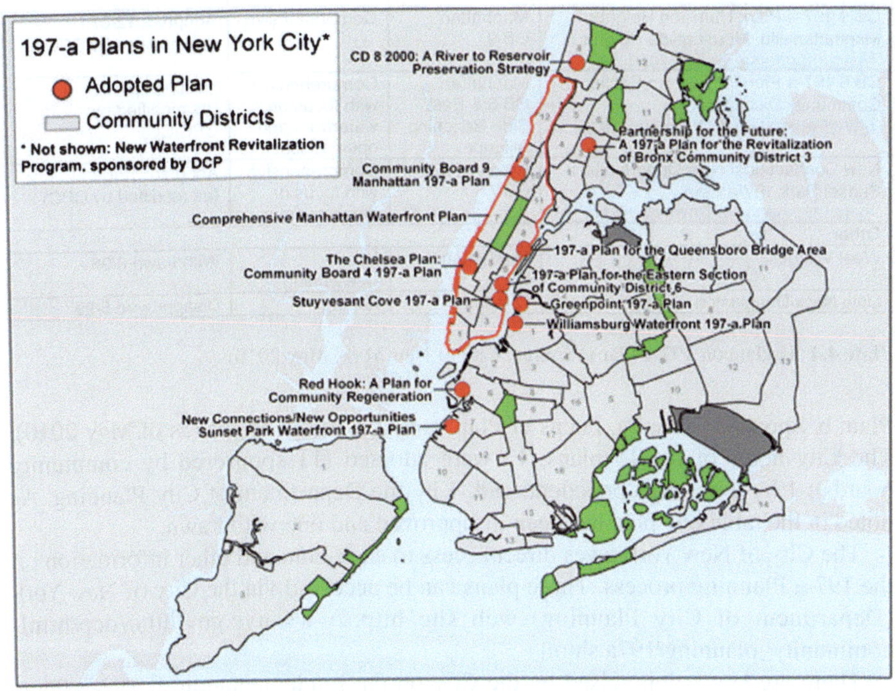

Plate 4.3 197-a Plans in New York City (Source: City of New York, Dept of City Planning 2013a, b)

Plan	Sponsor	Focus	Current Status
Adopted			
Partnership for the Future More Information & Download plan...	Bronx CB 3	Comprehensive	Adopted: 11/92 (as modified by CPC)
The Chelsea Plan More Information & Download plan...	Manhattan CB 4	Zoning	Adopted: 5/96 (as modified by CPC)
Red Hook Plan More Information & Download plan...	Brooklyn CB 6	Comprehensive	Adopted: 9/96 (as modified by CPC)
Stuyvesant Cove Plan More Information & Download plan...	Manhattan CB 6	Waterfront	Adopted: 3/97 (as modified by CPC)
Comprehensive Manhattan Waterfront Plan More Information & Download plan...	Manhattan BP	Waterfront	Adopted: 4/97 (as modified by CPC/CC)
New Waterfront Revitalization Program More Information & Download plan...	DCP	Waterfront	Adopted: 10/99
Williamsburg Waterfront Plan More Information & Download plan...	Brooklyn CB 1	Waterfront/ Comprehensive	Adopted: 1/02 (as modified by CPC)
Greenpoint Plan More Information & Download plan...	Brooklyn CB 1	Comprehensive	Adopted: 1/02 (as modified by CPC)
CD 8: River to Reservoir Preservation Strategy More Information & Download plan...	Bronx CB 8	Comprehensive	Adopted: 11/03
CB 8 197-a Plan for Queensboro Bridge Area More Information & Download plan...	Manhattan CB 8	Waterfront / Streetscapes	Adopted: 8/06
CB 9 197-a Plan: Hamilton Heights, Manhattanville, Morningside Heights More Information & Download plan...	Manhattan CB 9	Comprehensive	Adopted: 12/07 (as modified by CPC)
CB 6 197-a Plan for Eastern Section of Community District 6 More Information & Download plan...	Manhattan CB 6 & East Side Rezoning Alliance	Comprehensive with focus on waterfront and open space	Adopted: 3/08 (as modified by CPC/CC)
New Connections/ New Opportunities: Sunset Park 197-a Plan More Information & Download plan...	Brooklyn CB 7	Comprehensive with focus on the waterfront	Adopted: 12/09 (as modified by CPC)
Other			
West Village	Manhattan CB 2	Land Use/Zoning	Withdrawn 8/96
Little Neck/Douglaston	Queens CB 11	Zoning	Disapproved 5/99

Plate 4.4 Update on 197-a Plans (Source: City of New York, May 2010)

Plan; b. Sponsor Group; c. Focus of Plan; and, d. Current Status (as of May 2010). The City notes of the 15 plans, 13 were adopted (11 sponsored by community boards); 1 by a borough president; and, 1 by the Department of City Planning. As noted in the table one plan has been disapproved and one withdrawn.

The City of New York gives direct access to each Plan and other information on the 197-a Planning process. These plans can be accessed via the City of New York (Department of City Planning) web site http://www.nyc.gov/html/dcp/html/community_planning/197a.shtml

There are key points raised by the City on the implementation of 197-a Plans (City of New York 2013a) (www.home.nyc.gov). The sponsor of a plan, for example, (usually a Community Board) needs to work with city agencies in putting together and implementing a 197-a plan. In the instance where land re-zonings are recommended in the plan, the sponsor may encourage the Department of City Planning to initiate a zoning map change application.

The City notes in the Handbook that some 197-a plans have recommendations that focus on issues other than zoning. The first adopted 197-a plan, sponsored by Bronx CB 3, aimed at revitalizing the district and recommended measures to facilitate new mixed income housing development and increase the population. The City states that those goals have been substantially met.

The Stuyvesant Cove 197-a Plan envisioned a publicly-accessible waterfront park and pedestrian esplanade (the waterfront park was opened in 2003). A major recommendation of the Manhattan CB 8 197-a Plan was the transformation of a former heliport site to a waterfront park and esplanade. The City notes that (in consultation with Community Board 8), the City is planning these waterfront improvements.

The City has pointed out that the city's growing population (and strong real estate market) in 2013 has created interest in private or institutional redevelopment of under-utilized areas of the city. The City notes that in cases where rezoning is required, these proposals may be in conflict with community plans in various stages of development. The City thus urges local stakeholders (i.e. developers and Community Boards) to find common ground around the subject proposal.

In cases where there is a 197-a plan and a conflicting rezoning proposal, the City seeks to ensure that the competing plans are reviewed in a manner that guarantees equal consideration of each. The City presented these two case studies of resolving conflicts between the community and development proponents (City of New York 2013a, b) (www.home.nyc.gov).

> Start: taken from web (13 Nov 2013).
> Case Study 1.
> Columbia University proposed an expansion of its academic campus and other rezonings for one of the geographic areas covered by the Manhattan Community Board 9 197-a Plan. Public review began in June 2007 for Columbia University's expansion proposal in West Harlem and the 197-a plan proposed by Manhattan Community Board 9. In December 2007, the City Council adopted the CB 9 197-a Plan, as modified by the City Planning Commission, and the Columbia University rezoning proposal, as modified by the City Planning Commission. The recommendations in both plans, as modified, were reconciled.
> Case Study 2.
> The East River Realty Company (ERRC) proposed a rezoning for one of the geographic areas covered by the 197-a Plan submitted by Manhattan Community Board 6. Public review began in August 2007 for the East River Realty Company's proposal to redevelop the former Con Edison sites on First Avenue on Manhattan's east side. On January 28, 2008 the City Planning Commission approved the CB 6 197-a Plan with modifications, and the ERRC proposal with modifications. On March 26, 2008 the City Council adopted the CB 6 197-a Plan with additional modifications, and the ERRC proposal, also with additional modifications.
> End: taken from web (13 Nov 2013)

The City notes that community-based planning is essential to the city's vitality. People who are close to neighborhood issues, the City states, can clearly identify community needs and advocate passionately for local concerns. The City expresses its awareness that community-based planning comes in many forms. It can range from participation in local organizations to the preparation of a comprehensive community-based plan for official adoption. The City notes (City of New York 2013a, b) (www.home.nyc.gov):

> Community-based planning may seek to address a variety of issues including preserving neighborhood character, promoting affordable housing, facilitating new development and/or encouraging local employment. These goals may be pursued through rezonings, local plans and/or task force efforts. The Department of City Planning (DCP) provides technical assistance and advice to individuals and community-based organizations at all levels of planning.

The City states that community-based planning often begins at the Community Board level. The City directs these boards to its website. The web contains information on: data organized by community district (land use, population, housing, community facilities). The Department of City Planning website includes: an explanation of zoning; basic guide to New York City zoning; and, the Zoning Resolution (both text and maps).

Demographic information, data from the 2000 Census is provided by the City. The Census Fact Finder provides easy access to population information for a selected area.

Descriptions about land use and environmental review processes, and the status and details about Department of City Planning initiatives and other land use applications are also found on the website. In addition, the City offers a variety of data products, including base map files and land use data, for free download or by a license agreement.

The City asks community organizations to determine which community-based planning strategy is most appropriate for addressing any particular issue. The City outlines how problems can be addressed (including basic issues such as clogged drains, broken street lights, park maintenance problems, etc.). These may be dealt with at the community board's monthly District Service Cabinet meeting attended by representatives from city agencies (Police, Parks and Recreation, Sanitation, etc.).

The City urges sponsors of plans to note that taking a 197-a plan from inception to adoption is a lengthy process and requires the continuing commitment of its sponsors (even after adoption to ensure successful implementation). The City concludes that with a commitment and appropriate objectives, a Community Board may find the 197-a process well worth the effort.

The City outlines three options a community group may wish to pursue to address broader issues. These options are outlined below.

> Start: taken from web (13 Nov 2013).
>
> Option One
>
> A local zoning proposal developed in collaboration between the community and City. If a community-based organization seeks to change permitted land uses and/or building scale or density in a particular area, then a proposal developed in collaboration between the community and the Department of City Planning may be appropriate. The Department's borough offices can provide technical assistance to community boards and civic associations exploring such zoning solutions. Communities sometimes conduct their own field surveys to develop and support these strategies and expedite the process. Most often, the Department conducts the analysis, files the rezoning application and prepares the environmental review

documents, at no cost to the community organization or the community board. There are many examples of DCP/community collaborative rezoning efforts.

Option Two

A 197-a Plan, is usually sponsored by a Community Board. Long range and complex development issues may call for a comprehensive planning approach to identify goals and prepare a planning framework to achieve them. A 197-a plan may be appropriate.

Option Three

Inter-agency/Community Action Strategy. This strategy utilizes a task force made up of local representatives, city agencies and elected officials.

End: taken from web (13 Nov 2013).

The City goes further to explain "that some topics, such as improved traffic or building code enforcement, do not lend themselves to a formal plan or report but are more appropriate for a concerted action strategy. These subjects might best be dealt with through a task force made up of representatives of the appropriate agencies, community groups and elected officials." (City of New York, Department of Planning 197-a Plans website). The City then gives seven (7) examples of successful task forces formed to develop solutions to pressing local issues, as detailed below.

The City states (example one) that the Hunts Point Vision Plan (South Bronx) was a comprehensive initiative, as developed in cooperation with business and community leaders, elected officials and City agencies. The plan aims at promoting a competitive business environment and sustainable community on the Hunts Point Peninsula. The City notes that following the Vision Plan's recommendations, the Department initiated zoning measures (as adopted by the City Council) to encourage the growth of the food industry sector and create a buffer between the manufacturing district and the adjacent residential neighborhood.

In response to the broad range of concerns (in example two) expressed by participants about the future of 125th Street in Harlem, the Mayor formed the 125th Street Interagency Working Group. The Department of City Planning was joined by representatives from the Economic Development Corporation (EDC) and several other city agencies including the Departments of Cultural Affairs, Transportation, Small Business Services, and Housing Preservation and Development (HPD). The City highlighted how this group worked together with the Advisory Committee to identify solutions for issues raised during the planning process. As a result a rezoning proposal was adopted by the City Council (April 2008) with follow-up measures ongoing.

The redevelopment plan for Stapleton in Staten Island (example three) (including the former Navy Homeport) encompassed the construction of an almost mile-long esplanade along New York Harbor. This project came about from recommendations made in 2004 by Mayor Michael Bloomberg's Task Force on Homeport Redevelopment. As a result the City's Economic Development Corporation is fostering development of 350 residential units; a banquet hall; waterfront restaurant; sports complex; ground-floor retail and farmers market; and, with potential for a major economic user such as a movie studio or office space. The project was started when the City budgeted $66 million for the public improvements, including the construction of the esplanade and open spaces.

The Mayor's Staten Island Growth Management Task Force (in example 4) (convened in 2003 in response to overdevelopment in the borough) made recommendations that have resulted in significant zoning changes and enforcement improvements. The City notes that following these measures, new construction conformed to more desirable patterns. As a result the number of new permits (for construction) was reduced to a rate compatible with Staten Island neighborhoods. In addition, other recommendations emerged from the Task Force, including comprehensive studies of the West and North Shores of the Island.

A Transportation Task Force (example three) was established to address one of Staten Island's most serious concerns. It was comprised of elected officials, City agencies, State transportation agencies, Community Board chairs and the Staten Island Chamber of Commerce. The City states that the Task Force worked at the Mayor's directive in 2006 to produce a short term action plan. The plan contained medium and long term recommendations that were presented to address transportation issues focusing on development patterns, roadways and highways, bridges and mass transit. In the latest announcement the City concludes that significant progress has been made by the task force.

During public review of the comprehensive redevelopment plan for Jamaica (in example six), the City Planning Director (Amanda Burden) brought City agency commissioners to community meetings in Jamaica to develop planning strategies. This action led to examining longstanding infrastructure issues much like the Staten Island task forces which tackled similar problems (outlined above).

In 2007 the Mayor presented a vision for the revitalization of Coney Island in Brooklyn (example seven). The Department of City Planning had been preparing a comprehensive plan for the area. The City states that in this type of planning various agencies are brought together to cooperate, including: the Economic Development Corporation; Department of Parks and Recreation; the Department of Housing Preservation and Development; the Mayor's Office of Environmental Coordination; the Department of Environmental Protection; the Department of Transportation; the Landmarks Preservation Commission; elected officials; and, the community. As a result of this effort the rezoning was adopted in July 2009.]

End: taken from web (13 Nov 2013)

4.4 Outline and Review of 197-a Planning Process

The City outlines in the *197-a Plan Technical Guide* (City of New York) (1997) the required steps to create a 197-a Plan. The first four (4) steps of the nine (9) process (called 'threshold review') are contained in Plate 4.5.

http://streetswiki.wikispaces.com/Community+Boards,+New+York+City

Start: taken from web (13 Nov 2013)

Step 1. Letter of Intent/Plan Preparation

A letter of intent helps the City to allocate resources and identifies a sponsor (i.e. community based group) for creating the plan. The local community board would have decided that the local issues lent themselves to a 197-a plan and the resources were available to produce the plan.

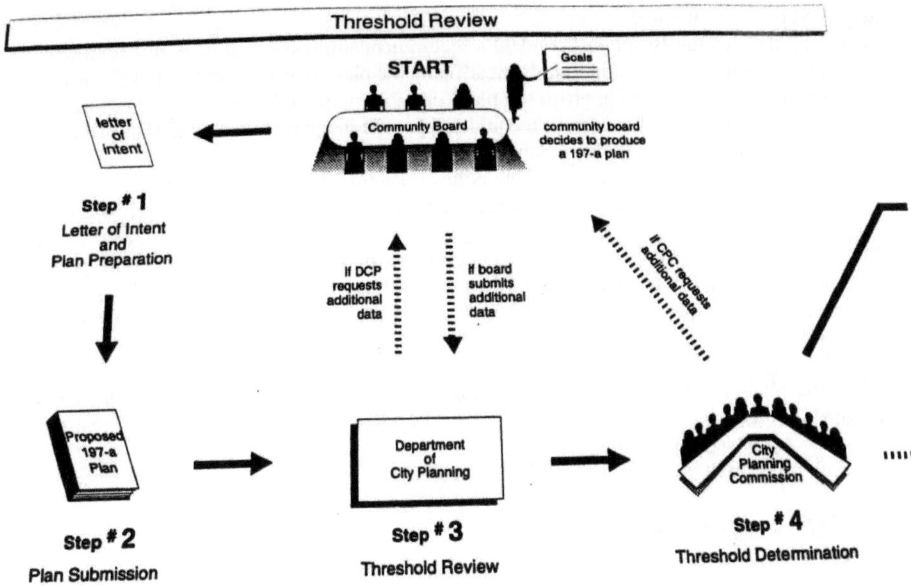

Plate 4.5 197-a Plan Threshold Review Steps 1–4 (Source: 197-a Plan Technical Guide 1997)

Step 2. Plan Submission

This step outlines how a completed plan is completed. The 197-a Guide outlines the importance of public participation in the formulation and preparation of the plan. The City wants a diverse range of viewpoints and conditions in the community to be acknowledged. The aim is for the community to achieve a consensus supporting the plan (i.e. via public forums and keeping major property owners within the study area informed about proposals that may affect them). A plan can take note of desired federal or state actions relating to the plan's key objectives. An agency so engaged needs to state its willingness to entertain the proposal. Finally, the plan needs to be consistent with City Council policies; or if not to state the reasons for the differences.

Step 3. Threshold Review

The City's Department of City Planning (DCP) has up to 90 days in this step to review the plan and report to the Planning Commission on whether the plan meets the threshold standards for form, content and consistency with sound planning policy. The DCP may ask the sponsor for additional information and documentation to correct any deficiencies in the plan. If the sponsor does not agree, the DCP submits the plan unchanged for the Commission's threshold determination.

Step 4. Threshold Determination

Here the Commission has 30 days to determine if the plan has met threshold standards. If the plan fails to meet these the Commission refers the plan back to the sponsor with an explanation of the plan's deficiencies.

Step 5. Environmental Review

Once the plan receives threshold approval, the Commission either directs DCP to begin an environmental review (up to 180 days) of the plan or it may defer the review to consider related planing efforts or land use proposals.

Step 6. Community/Borough Review

Once the environmental review has been completed, the Department of City Planning circulates the plan as follows: the affected community board; Borough President, and borough board; to affected city agencies; and, to any other community or borough board upon written request.

Step 7. Substantive Review

Within 60 days of the Borough President's recommendations (or 120 days if a review is held) the Commission holds a public hearing on the plan. Within 60 days of the hearing, the Commission votes to approve the plan (possibly with modifications) and reports to the City Council. The Commission could disapprove the plan, however in that instance the Mayor can request the City Council to conduct a review of the plan. The Commission reviews a plan based on sound planning policy standards and the sponsor's understanding of external relationships (legal and practical factors and consequences of plan proposals).

Step 8. Review

The City Council in any review has 50 days to hold a public review and vote on the plan. The Council adopts the plan and can add modifications. The Council then sends the plan back to the Commission to agree or request a whole new review. Alternately the Council can decide not to adopt the plan.

Step 9. Distribution

Once the plan is adopted the City sends copies of the plan to affected agencies. The agencies use the plan as a guide for actions to be taken by those agencies. Plans are also made available to the public.

End: taken from web (13 Nov 2013)

4.5 Applying the 197-a Plan Process

The 197-a plan recommends strategies to address a range of concerns, including: land use; housing; economic development; environmental or social issues. Reference http://www.nyc.gov/html/dcp/html/community_planning/presentation.shtml.

The 197-a Plan can be a community's master plan (wide range of concerns) or it may focus on a single issue (i.e. waterfront or zoning policies). The plan can lay groundwork for subsequent zoning changes or call for new municipal facilities or services (e.g. from day care to traffic measures). The City requests that when a request is made for a 197-a plan that the sponsor consider key points to be contained within the plan, including: plan's objectives; census tracts; area's land use and housing patterns, traditional neighborhood borders; zoning and designations such as historic areas; major roadways; institutions; parks; and, natural features. Plate 4.6 contains an example of a hypothetical study area showing prominent local features.

The 197-a Guide provides an example of a 197-a land use map (Plate 4.7). Here land use information can be used to: explore issues, particular problems; or, development or service opportunities.

The 197-a Guide concludes that if a community's vision is to be taken seriously, the plan must be logical, convincing and realistic. The City Council is responsible for providing the guidance and technical assistance throughout the plan process to facilitate a successful outcome for the 197-a plan.

Within Brooklyn for example in 2007 there were City adopted plans covering (Plate 4.8): Williamsburg Waterfront; Greenpoint; Red Hook; Sunset Park; the Bushwick Rheingold Site; Gowanus Estuary; Fifth Ave Housing Plan; Myrtle

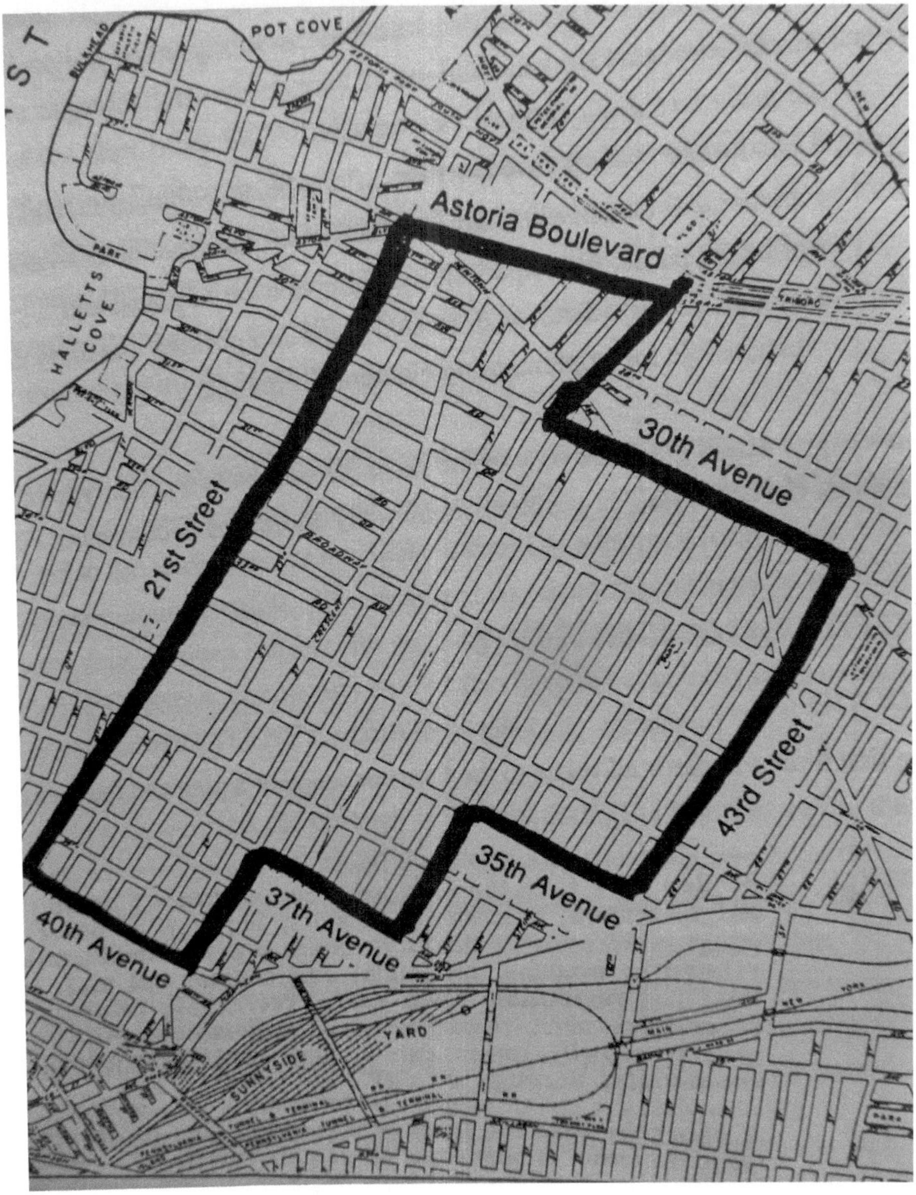

Plate 4.6 197-a Plan hypothetical study area (Source: 197-a Plan Technical Guide 1997)

Ave Revitalization in Fort Green; Old Brooklyn District; Brooklyn Bridge Park; Bedford-Stuyvesant; Coney Island Vision Plan; and Brooklyn Waterfront Greenway.

A sample of 197-a Plan locations within northern and western Brooklyn is contained in Plate 4.9.

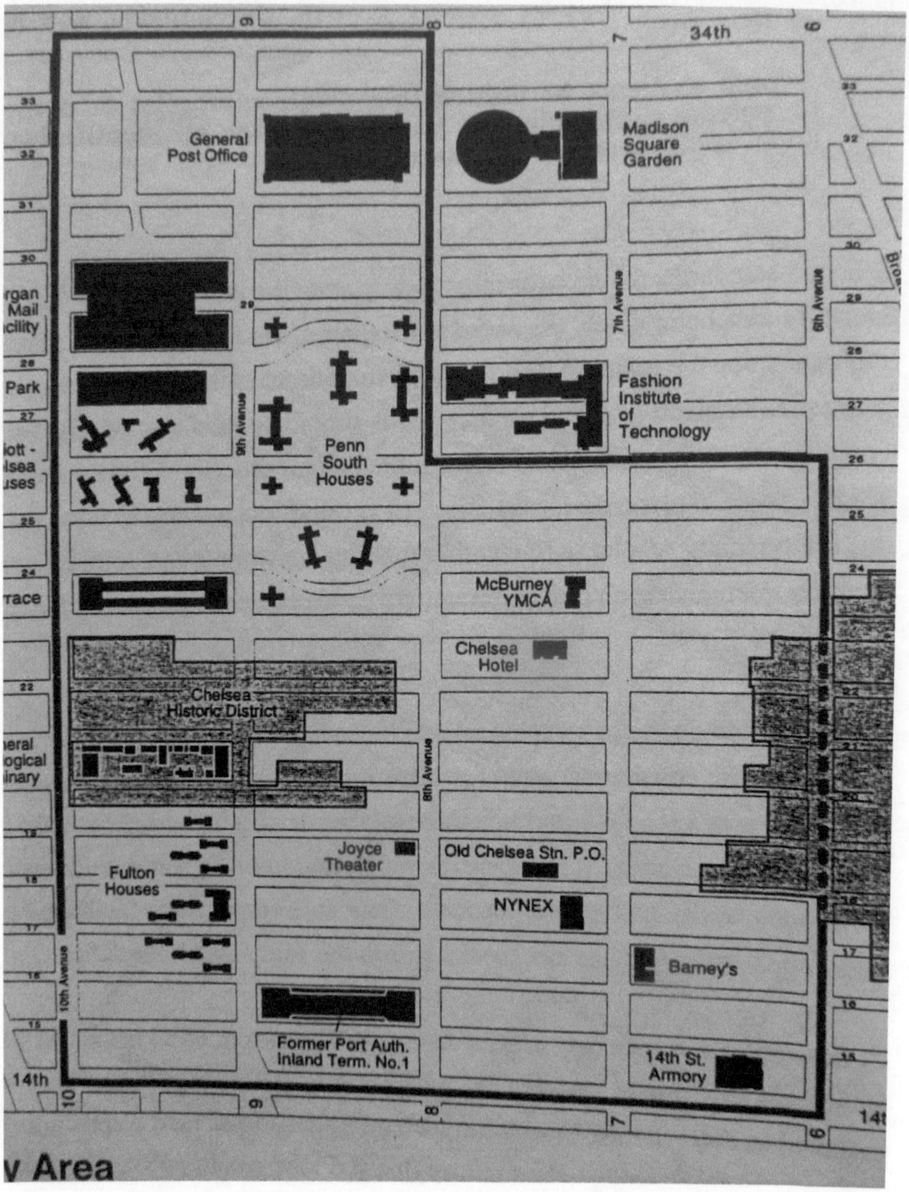

Plate 4.7 Sample 197-a Land Use Map, Hostos Community College Area Study, DCP, 1994 (Source: 197-a Plan Technical Guide 1997)

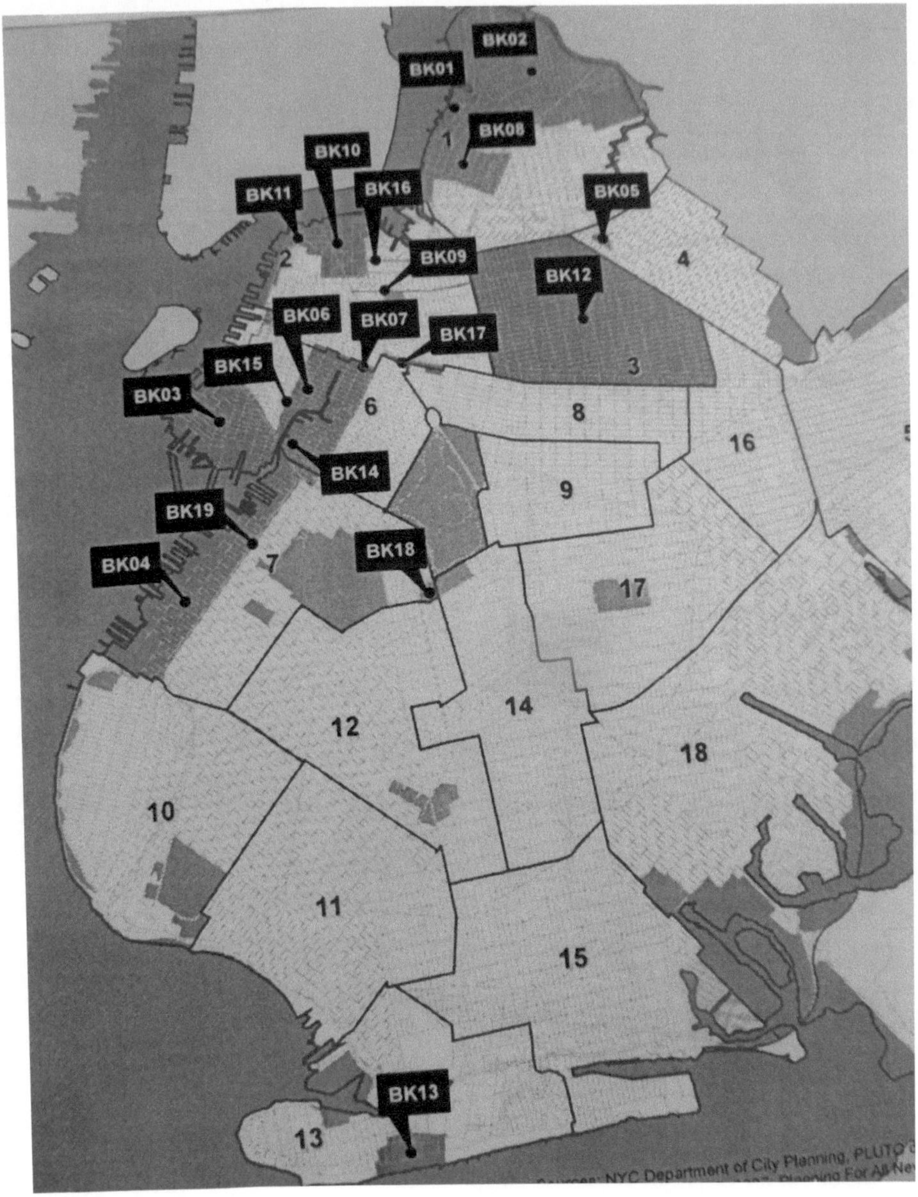

Plate 4.8 Brooklyn Community Based Plans by District (Source: Municipal Art Society Planning Centre 2008)

Plate 4.9 Northern and Western Sections of Brooklyn 197-a Plan Areas 2008 (Source: Municipal Art Society Planning Centre 2008)

4.6 Review of 197-a Planning

The Planning Centre of the Municipal Art Society of New York undertook a review of the 197-a Charter and its implementation in 1998 (Municipal Arts Centre of NY 1998). At the time the Centre noted it was too early to determine the long-term effect of the 197-a plans on NYC neighborhoods. The Centre however concluded there were some problems in the 197-a planning process that needed attention. The Society, as a result recommended: greater allocation of City Council resources and expanded dialogue with community interests; and, improvement in public sector (City and agencies) interest in the 197-a process. The Society went on to outline many of the benefits of the 197-a planning, including; a useful land-use planning tool; and, a community building mechanism.

The Society highlighted the community-based planning that had expanded across the United States through the 1990s, thus providing alternatives to traditional "top-down or development controlled planning". The Society was impressed that the 197-a planning process had generally established "broad community participation, collaborative partnerships, and strengthened local community capacities". Finally, the Society, in composing the review, compared 197-a planning with planning initiatives and neighborhood planning approaches of other cities in the United States. The report noted the number of cities that embraced comprehensive community based planning as a model for: funding and service delivery; and institutionalizing planning practice into local laws and ordinances. New regulations, for example, could direct city planning and community development agencies to enter into partnership with communities to develop comprehensive plans.

Other Society recommendations to the City Council included need for: improved working partnership with local communities; identifying a planning team with wide representation to develop specific recommendations within the plan; City assigning trained staff to assist in the 197-a plan making; and to tie 197-a plans more closely to the agency budgets and service statement, while promoting inter-agency collaboration.

While the City has pressed ahead with the 197-a Plan process, including consideration of the Municipal Arts Society recommendations, useful comments on New York City neighborhood planning and citizen involvement are offered by Prof. Tom Angotti (1999) in a journal article (http://www.newvillage.net/Journal/Issue1/1angotti.html). While the article was written over 10 years ago, planning events in New York City since then reinforce and in many instances validate the projected concerns.

The article as presented by Angotti is based on considerable experience gained by Angotti in the area of neighborhood planning. He is currently a professor at Hunter College (Manhattan) and has been professor and chairperson of the Brooklyn based Pratt Institute Graduate Center for Planning and the Environment. He has also been Executive Editor of Planners Network and Associate Editor of Planning Practice and Research. In addition, he also has had face-to-face experience with 197-a Plans, including as technical advisor to the Red Hook (Brooklyn) community plan. There are a number of key points Angotti raises on the subject.

In looking back, Angotti notes that neighborhood planning began with protest and organizing at the grass roots. He states that although there is an extensive official structure for neighborhood planning, many neighborhoods, especially low-income neighborhoods and communities of color, must still exercise their influence through protest and organizing. His research led him to the conclusion that the most salient issues (in 1999) were: waste transfer stations; community gardens; auto-dependence; and, access to public space.

In 1999 (with a city population of 7.5 million people and over 100 neighborhoods) Angotti argues that the official City of New York structure for neighborhood planning is honored 'more in the breach than in practice'. He states that "the official structure also leaves uncorrected the substantial political and economic inequalities among neighborhoods". He emphasizes that because better-off neighborhoods are usually able to show more muscle, their influence is greater, and their methods for organizing more discreet.

While the 197-a planning process has produced many good results, Angotti points out that the plans (like most official plans) can at best have a limited impact on neighborhoods. He states that "the plan is a policy statement and advisory". The plan, he notes, only obligates city agencies to consider the plan recommendations in making future decisions. He further states that the main problem is that when plans mandate narrow actions these actions can be ignored or changed.

The problem, Angotti believes, is that city government does not support or promote community planning in general, and 197-a plans in particular. He notes that the average community board covers an area of 100,000 people with a staff of two or three people who spend most of their time dealing with minor complaints ranging from potholes to traffic lights. The boards don't receive funds for planning, nor does the City provide planners to work for them. Angotti states that when the few communities that complete plans send them to the Department of City Planning for approval, the groups "find themselves subjected to extensive scrutiny and may have changes imposed on them without community review".

Going further, Angotti goes on to note that "the real success stories of community-based planning are more likely to be found in the many unofficial movements, particularly in neighborhoods whose official institutions have less access to decision-making power—low-income neighborhoods and communities of color. These include grassroots efforts in housing, open space, environment, public health and community services."

Participatory planning started with community organizing, Angotti states and notes that in the 1930s New York City was the site of militant tenant actions linked with the unemployed and labor movements. He provides a historical note of Marshalls evicting tenants and moving all their furniture onto the streets, neighbors organizing to move it back in again. After the War, tenants were able to secure the continuation of war-time rent controls. Angotti notes that this militancy in the post-War era included the Harlem rent strike in the 1950s and the Co-op City rent strike in the 1970s, and extensive squatting and homesteading.

Looking at the housing needs, Angotti fills in the historical gap stating that "a strong housing movement has helped make New York the U.S. city with the largest stock of rental housing, limited-equity c-ops, public housing, and the largest stock of municipally-owned housing." He states that about 15 % of all housing units in the city are still protected by some form of public or social ownership. He notes further that deregulation and privatization (spurred by the current real estate boom and an intense wave of gentrification) are making this stock unaffordable for people with low and moderate incomes. "In addition widespread abandonment has destroyed significant amounts of private rental housing" (he notes).

On another historical note. Angotti states that in the 1960s and 1970s, the largest community-based organizations were started by squatters and tenants in city-owned buildings. The massive neighborhood abandonment in the 1970s, he notes, left 150,000 private dwelling units in the hands of city government. Due to the flight of industry and white workers to the suburbs, redlining by banks and insurance companies, and the official neglect of neighborhoods occupied by people of color, landlords either torched their buildings for the insurance money or abandoned them completely, he states. The situation in Bushwick leading up to the fires of 1977 (as covered in Chap. 2) was fueled by this neglect.

On a positive note, under pressure from communities (and the real estate industry), Angotti notes that:

> New York City launched the largest-ever municipal housing program in the nation, resulting in the improvement of the vast majority of city-owned buildings. These efforts fed the creation of more than 100 community development corporations, neighborhood-based developers that own and/or manage over 10,000 units of housing. The city's many innovative programs that encourage tenant management and involvement, however, have now been eclipsed by efforts to sell everything to the highest bidder.

One phenomenon that Angotti draws attention to is that neighborhood solidarity had been gradually eroded by the enticements of gentrification, and community development corporations had too often lost sight of their service mandate and act like landlords. Therefore, in 1999, the housing movement was in a difficult place, He notes, "as indicated by a recent, nearly successful effort by the real estate industry to eliminate rent regulations in the city."

A historical perspective is taken by Angotti, writing that "in the predominantly middle class neighborhoods, militant community organizing helped lay to rest the spirit of Robert Moses, the planner responsible for massive urban renewal and highway programs that displaced and divided neighborhoods." The fight to stop the Lower Manhattan Expressway, and the 25-year-long struggle against Westway (a major highway on Manhattan's west side), he states, impressed city planners with the power of community opposition.

On a final note on community-based organizations, Angotti notes that not all of them have helped in the move toward more equitable policies. In fact, many a neighborhood and block association, especially in the more affluent areas, work overtime to preserve their territorial privileges, he states. With a sense of humor he notes "these are the neighborhoods for whom NIMBY (Not In My Back Yard) is a vocation".

In summing up, Angotti states that the most important contributions to community planning come from efforts to improve the environment in communities that have the most serious public health and environmental problems. In the 1980s, he gives an example, the City proposed to built a sewage treatment plant in West Harlem after real estate interests in the adjacent (mostly white) Upper West Side neighborhood successfully defeated the proposal. He goes further in outlining how Harlem residents organized but were unable to stop the plant, winning some concessions, notably a new skate park built on top of the sewage plant. Out of this struggle and similar ones, he states, against waste transfer stations, polluting industries and heavy truck traffic in the city's poor neighborhoods and communities of color, has arisen the environmental justice movement in New York City. He notes this movement is as critical to urban justice as the housing movement was in the 1960s and 1970s.

Taking a historical perspective, Angotti asks what has happened in recent decades that spurred this new movement of environmental concerns? He answers:

> First of all, the city has undergone a massive process of deindustrialization. The neighborhoods around the derelict, industrially zoned land tend to be disproportionately working class and minority neighborhoods. Historically, people in these neighborhoods have suffered the worst consequences of industrial pollution. They have been exposed to toxic substances both as workers inside factories and as neighbors living near them. Now much of the land vacated by industry is being taken over by waste facilities, which perpetuate the historic patterns of environmental injustice. Add to this the problems attendant to poverty and discrimination, such as lack of access to adequate health care, housing and education, and we have high rates of disease and infant mortality, as well as lower life expectancy in these neighborhoods. The central issue feeding demands for community planning in New York City today is waste disposal. Several of the city's poorest neighborhoods are saturated with waste transfer stations, which bring with them unsanitary conditions, heavy truck traffic, air pollution, and odor pollution.

Concluding on a positive historic note, Angotti states "that the community movements have done more in the long run to improve the quality of life for the greatest number of people in the city. They are working for a cleaner environment, public places and an end to environmental inequities. They, more than government, are advocates of planning – participatory, democratic planning."

A further review of the 197-a Plan program was undertaken by Todd Bressi within *Planning* (magazine) (2000). Bressi, at the time, was the executive editor of the design journal Places (based at the Pratt Institute, New York City). He raises questions of: how effective the 197-a Plan program has been in terms of expenses, time-consumption and complications of community planning. Bressi reflects on the history of community planning by noting "it was in the 1960s that Senators Robert Kennedy and Jacob Javits who helped launch the community-based development movement with the creation of the Bedford-Stuyvesant Restoration Corporation. At the same time, he notes, Pratt Institute opened its Center for Planning and Environmental Development, making university resources available to the neighborhoods." To the criticism that the 197-a process is too slow and the impact of plans on city policies is meager, defenders, he notes, "counter that the process is a way of drawing a broad spectrum of local residents into planning discussions to test their proposals in the political arena." Bressi points to shortcomings of the 197-a process in area of: changes resulting from City's reluctance to step on "agencies" toes; and, the plan is only advisory, with the City not always a willing partner. A valuable comment that the City often makes, Bressi points out, is that it is vital the community (through the 'sponsors' of the plan) to keep the ball rolling after the plan is complete via budget, land use and political processes. In addition, the City points out, a community board can use the plan as soon as it's done to review land use applications and set priorities for the district on an annual basis.

Bressi makes other valuable comments on the planning processes available in New York City. He notes there are alternative plan making processes the community can engage in. These include normal zoning and urban renewal processes, or to draw up its own plans. Another possible route for a community to (CCRP). This program, he points out was initially use to plan six Bronx neighborhoods. He explains, in examining the CCRP planning process, that these plans "address the quality of life" gaps (in parks, shopping, education, youth services, and employment) created by the City's housing redevelopment programs in the 1980s. An advantage of the CCRP plans is that, Bressi quotes a planner, "they are implementation oriented". A key component with CCRP, says a director of the program, is the investment of local community development corporations (CDC) (offering experience, political understandings and staff resources). The director also noted that charitable foundations can provide seed money to get plans under any planning process started. Bressie concludes that it would be advantageous for Community Boards to work more closely with community development corporations (CDCs).

The MAS produced the *Planning for All New Yorkers: An Atlas of Community Based Plans in New York City Campaign* (Municipal Arts Society 2008). The atlas represents the efforts of many grassroots organizations to present the City and their communities with plans for needed improvements. The Atlas (containing 104 community-based plans), MAS points out, is the only publicly accessible compilation of the City's community-based plans. Advances of the Atlas, MAS notes, include: prodding local candidates running for office an overview of planning work already done by communities; a tool for communities to a plan but do not know where to start; identifying common themes of plan making that communities face; and providing a living document that is updated on an ongoing basis as new plans are realized.

The MAS notes, on the plan production challenges, "community-based plans represent among the best planning being done in New York City…communities have turned to foundations, banks and technical assistance providers for support in development their plans". The MAS comments however that "plans are often adopted by the City and then left unused or unimplemented. City agencies sometimes regard community planning…as separate (even conflicting) of interests (with agencies)". The MAS points out that, only after many years of effort did some success appear, including: getting agencies to share pubic data; convincing City and State agencies to accept innovative proposals as valid and worthy of inclusion in a plan; organizing …to get plans adopted that may not coincide with market pressures; stopping other plans and proposals being implemented before community plans have been adopted; securing implementation funding; and lobbying decision makers to gain capital and expense investments, land use and zoning proposals and approvals (that a plan would deem important). Commenting on the 197-a process, the MAS concludes the goal is to establish community-based planning as official New York City policy. This would mean for the City to go further than the 1975 City legislation that allowed Community Boards draftee master plans.

4.7 Chapter Summary

This chapter examined the structure of community boards and planning instruments in New York City. The history of the creation of community boards in the City was summarized first. A critique of these boards was then offered. The chapter also examined the range of planning instruments available to the City of New York. Focus is mainly placed on the brownfield development planning and 197-a Plan process. With this understanding of planning tools and community engagement structures, the next chapter reviews and critique how planning for urban renewal is applied in the City of New York, using Bushwick and other parts of Brooklyn as case examples.

References

Angotti T (1999) New village. http://www.newvillage.net/Journal/Issue1/1angotti.html/
Bressi T (2000) Planning journal. American Planning Association, New York
City of New York (2013a) Guidelines for 197-a Plans. City of New York, New York. (www.nyc.gov)
City of New York (2013b) Handbook for Community Boards. City of New York, New York. (www.nyc.gov).
Municipal Arts Society (2008) Planning for all New Yorkers: an Atlas of Community Based Plans in New York City Campaign. Municipal Arts Society, New York

Chapter 5
Urban Renewal and 197-a Plans: Bushwick and North East Brooklyn

5.1 Introduction

This chapter looks at community boards in New York City and the application of planning instruments, especially the 197-a Plan process. The chapter focuses on Bushwick, Williamsburg, Greenpoint, Downtown Brooklyn and South East Brooklyn. The Bushwick planning example centers on the use of the Brownfield Redevelopment Plan program, applied to the West Bushwick area (Rheingold Renewal Project). The 197-a Plan applications to Williamsburg Waterfront and to Greenpoint are examined next, including historical backgrounds of both these North East Brooklyn neighborhoods. The planning of both Williamsburg and Greenpoint under the 197-a Plan process is also examined. To move beyond the Northeast in examining other planning instruments (such as master planning), two other neighborhoods are examined: Brooklyn Downtown and South East Brooklyn (including Coney Island).

5.2 Bushwick and Brownfield Redevelopment Plan

Through the efforts of a number of groups and the City of New York a West Bushwick renewal project (Rheingold Renewal Project) got underway in early 2000. This project comes under the City of New York's Brownfield Redevelopment Plan program (www.home.nyc.gov).

At the time (Bushwick Observer 2000) the United States Environmental Protection Agency estimated there were 450,000 vacant brownfields across the country. The article goes on to state there are thousands of acres of brownfields in New York, mostly in waterfront and industrial areas. It is the clean-up costs that stop developers from considering developments of these sites. It was here at the Rheingold Brewery site in Bushwick that the City and State could see a proposal to create a model of revitalization. The State member, Vito Lopez, in looking at the

R.C. Rauscher and S. Momtaz, *Brooklyn's Bushwick - Urban Renewal in New York, USA: Community, Planning and Sustainable Environments*, DOI 10.1007/978-3-319-05762-0_5,

Rheingold site proposed a bill to the State Legislature to create an advisory panel to maintain clear clean-up standards for any brownfield site in the State by setting clear clean-up standards. The site was once not only a brewery, but also: a gas station, a garage, a waste paper plant, auto repair facility and food products factory (thus potential contamination of the site from these uses).

The Ridgewood Bushwick senior Citizens Council Inc (RBSCC) was an initial force to see the potential renewal of the site for housing and social needs of the West Bushwick community. The article points to the wider challenges of renewing brownfield sites around the world with the setting up of the International Brownfield Exchange (IBE). The IBE had been coordinating the exchange of ideas about innovative design, policy, financing and other aspects of regeneration of brownfield sites since the mid-1990s. The planning process included the preparation of a Land Use Review Application and an Environmental Impact Statement. Given the history of the settlement of Bushwick in the late 1880s by immigrants from Germany, it was a note of irony that the planners and architects that came to draw up plans to restore this tragic abandoned site would come from the same cities that their forebears had left to go the America. Those cities represented in the planning team from Germany included: Dessay, Bitterfeld, Dessau, Leipzig, Halle, Markkleeberg, and Wolfen.

The sub-division, called Bushwick Garden City, layout is illustrated in Plate 5.1.

The projects building position and landscaping of Bushwick Garden City building and landscaping plans is illustrated in Plate 5.2. The plan made allowances for: green space with new community facilities; reconnections to the neighborhood; a green buffer; renaissance of Bushwick Ave; a day care center; supermarket; housing ownership for low income people; and, rental housing.

The historic importance of the West Bushwick area had been acknowledged by the NYC planners, with restoration of most of the buildings now planned (Plate 5.3). These include (left to right): Arion Hall (1887); St Mark's Lutheran Church (1892); William Ulmer Brewery; and former PS 62 (Ellery St).

Some useful urban planning insights came out of The Campaign for Community Based Planning as organized by the Municipal Arts Society in the late 1990s and early 2000s. The Municipal Arts Society comments on the Rheingold Brewery Plan (within Brooklyn Community Board 4) are centered on the document *Planning for All New Yorkers: An Atlas of Community Based Plans in New York City Campaign* (Municipal Arts Society 2008). This document came out of the work of the Community-based Planning Task Force commencing in 2001 and coordinated by the Municipal Art Society Planning Centre (www.mas.org). The Task Force is a coalition of grassroots community organizations, community boards, civic groups, elected officials, environmental justice advocates, planning professionals, and academics. The Municipal Arts Society had two member famous writers on urban planning in the past that gave the MAS a sound community planning basis, Jane Jacobs and William Whyte (as noted earlier). The motto of the MAS continues to be implemented, 'the art of making New York livable'.

The Task Force, in an overview, notes that "the growth of the brewing industry (in West Bushwick) sparked the construction of new homes and churches in the neighborhood as well as a diverse business sector, some of which are still

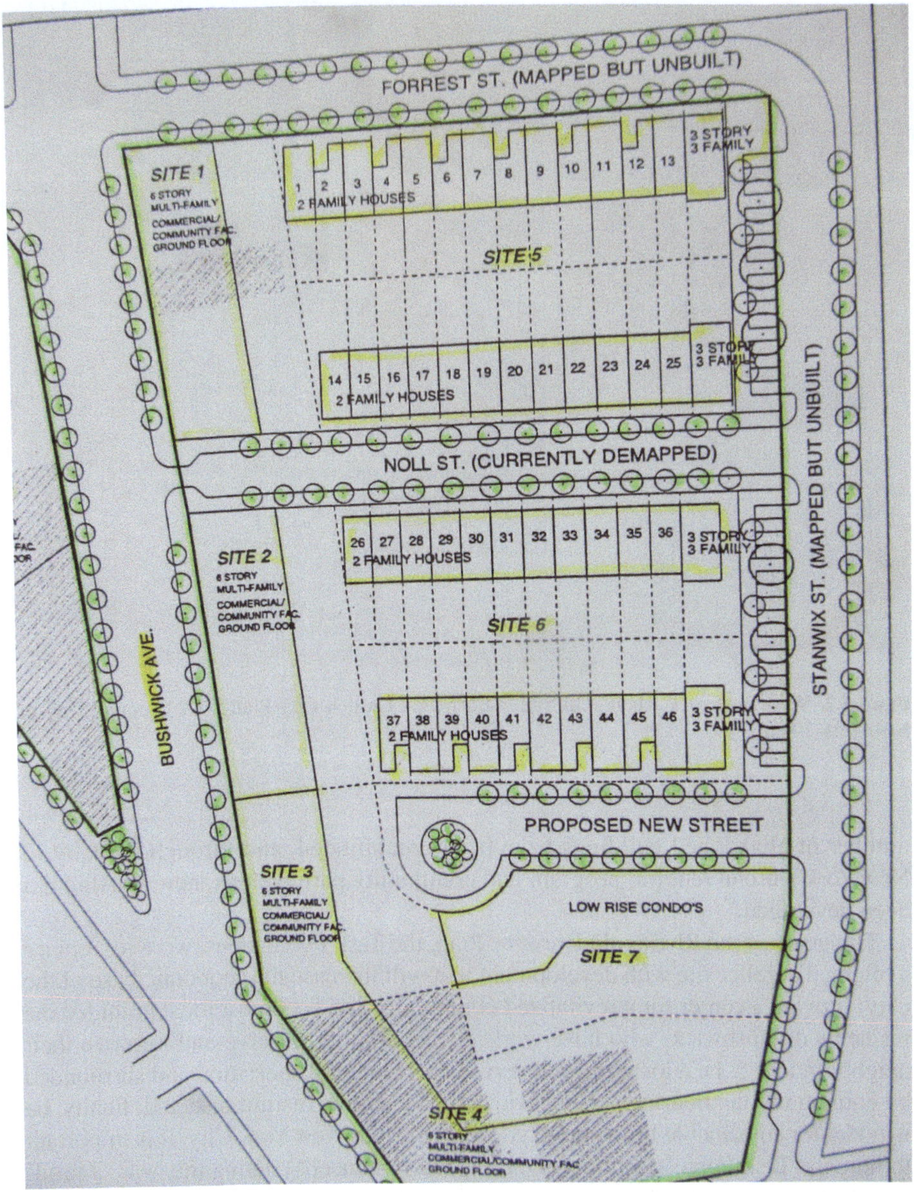

Plate 5.1 West Bushwick Rheingold Site – Project Planning Layout (Source: City of NY Layout 2002a, b)

prominent features of today's landscape, and created a vibrant community where people lived and worked." On a positive note the Task Force notes that "The community has put much effort over the years to improve surround conditions – a large

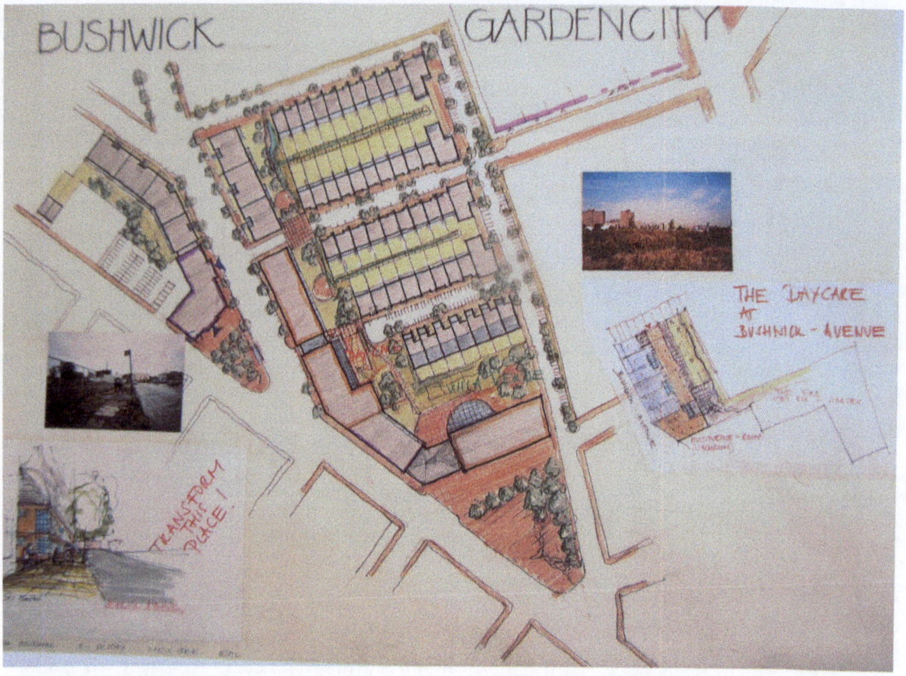

Plate 5.2 West Bushwick Rheingold Site – Bushwick Garden City Plan 2002 (Source: City of New York 2002a, b)

number of abandoned buildings have been rehabilitated, and through the City of New York's urban renewal program and community partnerships, new housing has been developed."

The goals of the Rheingold Brewery Plan, the Task Force notes, were to: replace a blighted derelict site with development that will increase the economic base of the City; provide a center for a revitalized community and become a focal point for the residents of Bushwick, who have worked diligently to preserve and improve their neighborhood; …be a location well served by public transportation and surrounded by community landmarks such as civic and educational institutions; and, finally, be a model for sustainable brownfield redevelopment in New York City. It is important to note the Task Force's comments on the design concept (stating this was 'a product of consensus and reflects the community's desire for action').

Some of the results, the Task Force notes, included: greenfield infrastructure can recreate physical and visual connections with the neighborhood (including a main public plaza with a commanding view of the Manhattan skyline); a green roof was installed over the community space; and, vacant portion of the site to be developed into a park.

Commenting on the housing, the Task Force, the plan provided: maximum number of units were proved while assuring diversity and pleasant environment(s); affordable

Plate 5.3 West Bushwick Historical Buildings. *Left* to *Right*: Arion Hall (1887); St Marks Lutheran church (1892); William Ulmer Brewery; and, Former PS 52 (Source: Environmental Study Rheingold Site, CNY 2000)

housing for a wide range of income groups; a mix of housing suitable for persons with varied lifestyles; and a maximum use of existing and successful housing programs (under the City's Housing Preservation Department). On the commercial space and community facilities the Task Force notes: space for neighborhood retail stores provided on the ground floor of multi-story buildings (6–8 stories) planned along Bushwick Ave. Two floors at Rheingold Gardens are occupied by the headquarters of a large social services agency (Ridgewood Bushwick Seniors Citizens Council) and a home health care agency and training facility; and, state of the art Senior Center.

Finally, the Task Force highlights the participatory process that the Rheingold Brewery Plan program promoted. An International design workshop (18–20 Oct 2000) was organized that brought together architects and urban planners from Germany, Northern Ireland and Canada (with experience in transforming brownfield sites into new uses). Attending the workshop were local community leaders, elected officials, city staff and professionals in the redevelopment field (see photo of international design team in Plate 5.1). The Task Force concluded that the workshop established consensus regarding future objectives and design options for the site. The Task Force, in observing the success of the planning, noted "Many people in the community are amazed at the change that (had) taken place, and (thanks to (the protocol of) 50 % community preference)) many of the people living on the site are longtime residents of the Bushwick community."

5.3 Williamsburg Waterfront 197-a Plan

Williamsburg Waterfront contains a rich past and remains a vital part of the North East Brooklyn area (Bushwick and Greenpoint being the remainder of this district). Some background notes on Williamsburg planning are summarized below (Department of City Planning, City of New York 2013) (www.home.nyc.gov).

The City points out that in the nineteenth century, Williamsburg, as it was then known, became a major industrial center, with notable capitalists like Cornelius Vanderbilt building grandiose mansions on the banks of the East River. Major corporations such as Pfizer Pharmaceutical, Domino Sugar and Dutch Mustard all got their start as Williamsburg factories.

The neighborhood's unique allure, the city notes, is its combination of towering brick factory buildings and charming, tree-lined side streets. On a poignant note, the City adds this Williamsburg scene is perhaps captured most memorably in Betty Smith's *A Tree Grows in Brooklyn*, a girl's coming-of-age story in the predominantly Irish, German, and Jewish neighborhood Williamsburg at the turn of the century. The City notes, in fact, the novel is set only several blocks from the present location of The Brooklyn Latin School (this school is commented on in Chap. 6).

Today, the City notes, the neighborhood reflects the city's rich and ever-changing tapestry: many of the factories have been converted into expansive lofts or art galleries, with Bedford Avenue being considered among New York's prime destinations.

At the same time, the social life in Williamsburg is noted by the City as the area "remains true to its legacy as a first home for new Americans, and it's easy to lose oneself among its ethnic eateries, sampling steaming Polish pierogies or authentic Dominican mofongo, among countless other delectables."

The city notes that since its heyday in the late nineteenth and early twentieth centuries as one of Brooklyn's foremost industrial hubs, DUMBO (Down Under Manhattan Bridge Overpass). This main section of Williamsburg has transitioned to a burgeoning, mixed-use community through the adaptive reuse of its historic loft and warehouse buildings with new light industrial, arts, commercial and residential uses.

Following the departure of its manufacturing base, in the 1970s and 1980s DUMBO became an attractive home to artists who found the large floor plate loft buildings ideal artisan workspaces. The City notes that "through both private and City initiatives over the last decade, mixed-use zoning regulations were introduced to the section of DUMBO between the Brooklyn and Manhattan Bridges. The mixed-use zoning encouraged private reinvestment that has resulted in the successful transformation of this area into a thriving, mixed-use neighborhood with a variety of shops, markets, offices and arts-related uses comingling aside a new residential population within DUMBO's rich and unique architectural fabric." The City notes that its Department of City Planning seeks to expand on the success of the mixed-use regulations that have fostered the resurgence of DUMBO to the section of the neighborhood east of the Manhattan Bridge.

In consultation with Community Board 2, local civic groups and elected officials and in coordination with the Landmarks Preservation Commission (LPC), the City developed a rezoning proposal for the section of DUMBO east of the Manhattan Bridge. This would allow residential conversion of existing loft buildings and foster new mixed-use construction while providing predictability and height limits that reflect the area's historic character, the City notes. It would also, for the first time in DUMBO, provide zoning incentives for the creation of affordable housing in new construction.

To those ends, Department of City Planning proposed to rezone existing zoning districts to mixed-use zoning districts that would better reflect and preserve the unique character of the neighborhood. In conjunction with the rezoning, the city notes, a related zoning text amendment would permit the use of the Inclusionary Housing Program which provides incentives and opportunities for the development of affordable housing in DUMBO. The proposal, the City points out, advances the past Mayor Bloomberg's sustainable planning goals by promoting the preservation of neighborhoods with special character while also providing opportunities for modest growth and affordable housing along wide corridors well served by mass transit.

In outlining the district's history, the City notes that DUMBO's emergence in the early nineteenth century as a bustling manufacturing hub resulted from the success of steam ferries that crossed the East River from the nearby Fulton Ferry Landing. The city states that "the development activity that was generated by this commerce had a profound impact on the neighborhood. Industrial companies began to develop the area with high density brick and reinforced concrete loft buildings to take advantage of the ease of storing, refining and shipping materials from this transportation hub. These loft buildings that continue to define DUMBO's character today, rise at the street line without setback and are punctuated with large windows, which allowed sunlight into the industrial workspaces. Following the general trend of industrial decline in Northeastern cities, the companies and workforce that operated from loft buildings in DUMBO gradually diminished by the middle of the 20th century."

Following land use and market trends of recent decades, the City points out that DUMBO has evolved from its industrial past into a dynamic, mixed use community that has adapted by housing a new generation of uses. A residential presence grew in the late 1970s as artists priced out of gentrifying Manhattan neighborhoods transformed the large spaces into work studios and residences. Beginning in the mid-1980s, the New York City Loft Board regulated the legalized conversion of many of these units as Interim Multiple Dwellings (IMD). The City concludes that since this time there has been a steady increase in residential occupancy within the existing loft buildings. The City notes this has occurred through both illegal conversions and Board of Standards and Appeals (BSA) variances.

Art galleries and local retail shops, among other design-oriented users, the City notes, have located predominantly in ground-floor spaces to service the increased activity in the neighborhood. The neighborhood's large footprint loft spaces and attractive rental prices, the City implies, have attracted creative commercial companies in new media fields such as web and design, music production and promotion, and graphic arts.

On a heritage note and in recognition of DUMBO's unique built character, the City notes that the neighborhood has received two historic district designations in recent years. The proposed rezoning area is completely within the DUMBO Industrial District established by the State and National Register of Historic Places in 2000, and is almost entirely within the DUMBO Historic Districts established by the Landmarks Preservation Conservancy in 2007.

Residents and developers want to know how rezoning affects them. On the note, the City states that "the proposed rezoning area contains approximately 12 blocks situated along Jay Street and Bridge Street, the neighborhood's two predominant north-south corridors. Buildings along the Jay Street corridor vary greatly in their respective size, ranging from 1 to 12 stories. The few low-rise buildings of 1 to 3 stories typically have ground floor uses of neighborhood services such as restaurants or local retail with upper floors containing residential use. Buildings of 3 to 7 stories rise from 50 to 80 feet, due to the tall ceiling heights of the loft structures. These are more likely to be uniformly occupied by warehousing or studio uses as several of these buildings have loading docks or other structural conditions that have prevented ground floor conversions. Lastly, the largest loft buildings of 6 to 12 stories, or 80 to 120 feet in height, are typically of mixed-use character. Although there are some vestiges of earlier light-manufacturing uses such as waste paper transfer, ground floor uses typically include dry cleaners, art galleries, garment production, and furniture sales. The remaining floors of these lofts contain an array of uses, including architectural, graphic design, and other business services, printing companies, woodworking and similar light industrial work, and art studios."

Buildings in the Bridge Street corridor, the City notes, commonly range from 4 to 7 stories, or 50 to 75 ft. The City adds that "smaller foundries and garages of 20 feet in height are also interspersed. Both of these building types do not typically have ground floor uses that interact with the street and are more likely to be occupied by industrial uses including furniture production, woodworking, and metal sculpting that have remained during the area's transformation. There are also concentrations of office space and residential units throughout the Bridge Street area as well."

In closing statements about the rezonings the City adds that the proposed action follows six previous private and Department of City Planning rezonings. These had occurred in the past 10 years in surrounding and adjacent areas that have increased the mixed-use character of the neighborhood. The City points out that these changes include: a. the mapping of a district in 1998 to allow for the mixed-use conversion of buildings along Main Street; b. the contextual residential rezoning of Vinegar Hill, in conjunction with the Landmarks Preservation Council designation of an historic district; c. the creation in 1999 of Brooklyn's first Special Mixed Use District, to allow for mixed-use conversion and new construction between the Manhattan and Brooklyn Bridges from Water to York Streets; and, d. several private rezoning applications to facilitate new mixed use developments. The City announced that these actions had generated over 900 new residential units with the capacity for creation of an additional 900 units in the future.

On the subject of affordable housing, the City notes that the related zoning amendment will make the Inclusionary Housing Program applicable in the proposed districts. This would establish incentives for the creation and preservation of affordable housing in conjunction with the development in the area. Under the Inclusionary Housing program, developments providing affordable housing are eligible for a floor area bonus, within contextual height and bulk regulations tailored to this area. Affordable units can be provided either on the same site as the development earning the bonus, or off-site either through new construction or preservation of existing affordable units. The City notes that off-site affordable units must be located within the same community district or within a half-mile of the development. Its noted that available City, State, and Federal housing finance programs may be used to finance affordable units. The City concluded that the combination of a zoning bonus with housing programs would establish an incentive for the development and preservation of affordable housing in DUMBO.

5.4 Greenpoint 197-a Planning

Further 197-a Planning in Northeast Brooklyn took place within the Greenpoint-Williamsburg Contextual Rezoning (2000) (Department of City Planning, City of New York 2013) (Plate 5.4) (www.home.nyc.gov).

Greenpoint is an older part of Northeast Brooklyn, with Williamsburg to the east and Bushwick to the south. Greenpoint has always been a working class area of Brooklyn, with easy access to work (especially along the foreshores of the East River).

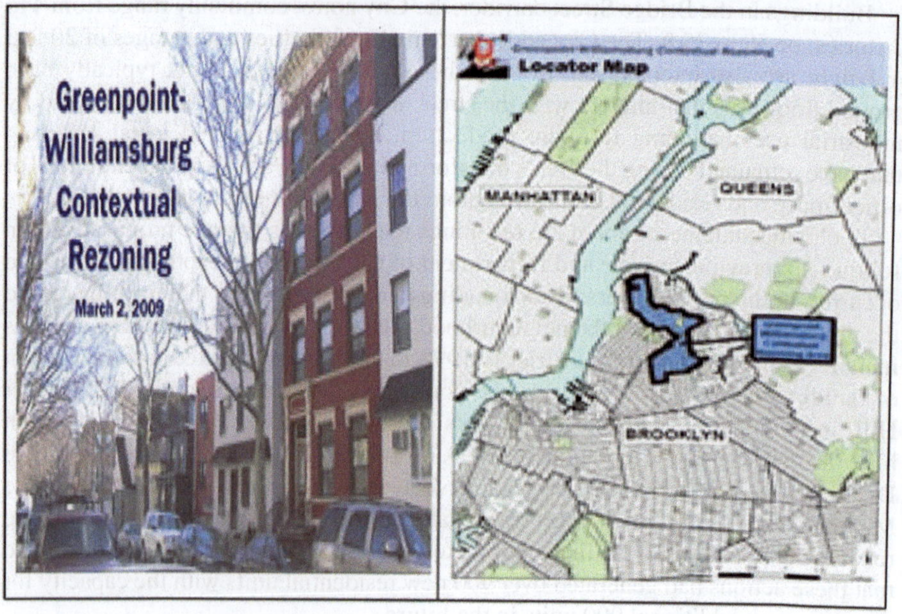

Plate 5.4 Greenpoint-Williamsburg Contextual Rezoning (Source: City of New York 2000)

The City points out that (www.home.nyc.gov):

> Greenpoint is largely working class and multi-generational; it is not uncommon to find three generations of family members living in this community. The neighborhood is sometimes referred to as Little Poland due to its large population of working-class Polish immigrants, reportedly the second largest concentration in the United States after Chicago. Although Polish immigrants and people of Polish descent are present in force, there is a significant Latino population living mostly north of Greenpoint Avenue, and Greenpoint has a significant number of South Asian and North African residents.

A brief look into the past of Greenpoint is revealed in Plate 5.5 (left to right): loading oil at the refinery, ca. 1950s Newtown Creek; oil refinery workers at quitting time, ca. 1950s; St Elias Greek rite Church; working families in Greenpoint, early 1900s; Dutch farmhouse (Duryea House) ca. 1800; and, Russian Orthodox Cathedral.

The social and economic life of Greenpoint and Williamsburg has always been buoyant. Plate 5.6 gives a glimpse of that life, left to right: McCarren Park Pool, 1936–37; the Monitor School ca. 1900; main street shopping in Manhattan St (1982); Greenpoint Savings Bank ca. 1930; Williamsburg summer day, Lynch St in early 1900s; corner shop in Williamsburg selling Jewish 'felty' hats.

The author's (Rauscher) field trips to Bushwick, Greenpoint, and Williamsburg (as noted in the Authors' Statement) occurred in 1979, 1982, 1993, 1998, 2000, 2007, 2008, and 2010. These trips enabled a closer research look at the urban renewal occurring in these neighborhoods (looked at earlier in this chapter). A snippet of those trips in Greenpoint and Williamsburg are illustrated in Plate 5.7.

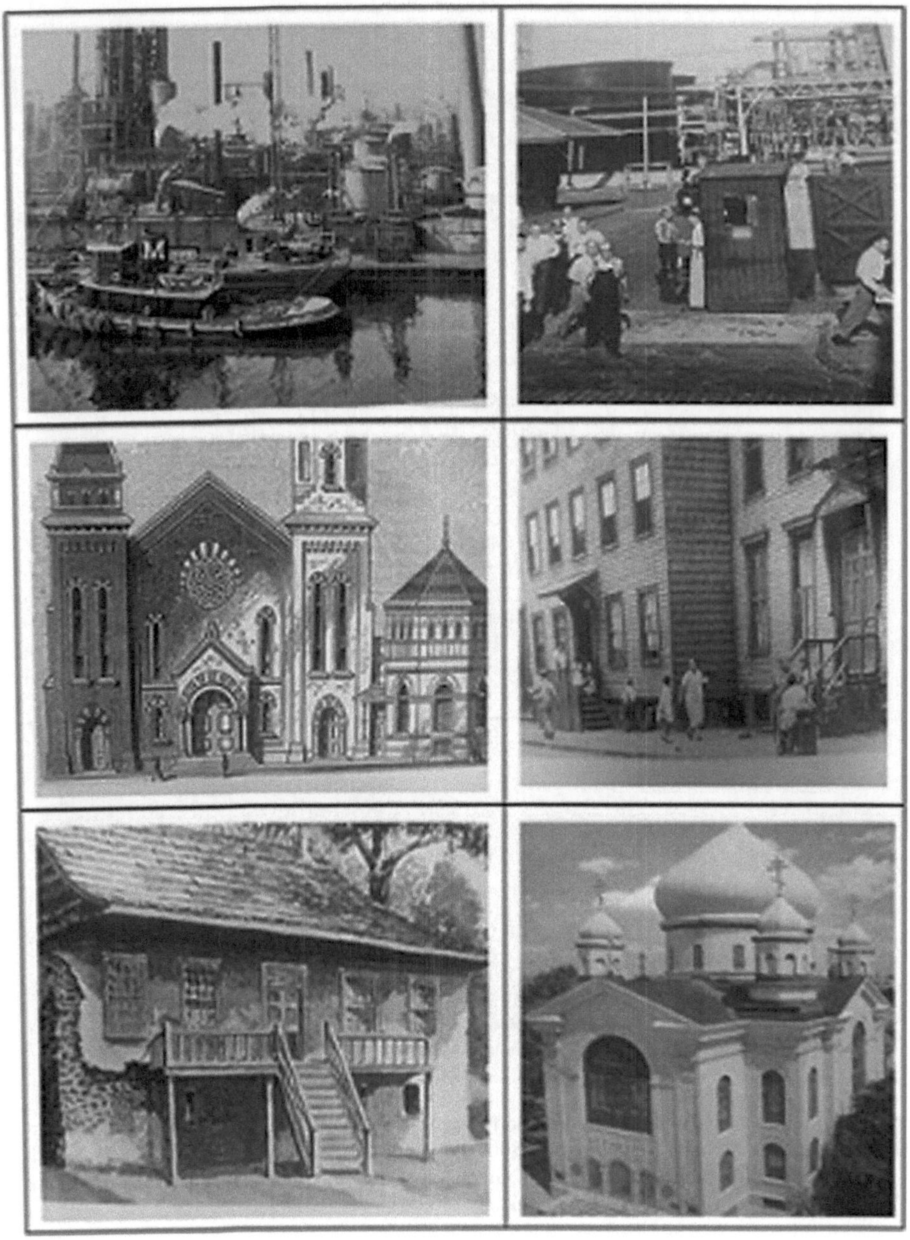

Plate 5.5 Greenpoint – Industry and Community Background. *Left* to *Right*: loading oil at the refinery, ca. 1950s Newtown Creek; oil refinery workers at quitting time, ca. 1950s; St Elias Greek Rite Church; working families in Greenpoint, early 1900s; Dutch farmhouse (Duryea House) ca. 1800; and, Russian Orthodox Cathedral (Source: Brooklyn Historical Society (2005))

Plate 5.6 Williamsburg and Greenpoint – Early Transport and Services. *Left* to *right*: Williamsburg-Water Transport and Sugar Refineries ca. 1800s; Wallabout Markets ca.1900s; Summer day Lynch St ca. 1900s; the Monitor School ca. 1900; Greenpoint – Savings Bank ca. 1930; McCareen Park Pool, 1936 (Source: Brooklyn Historical Society (2005))

Locations here include: the Manhattan St (Greenpoint) shopping street. The street continues to successfully trade given the stable population and industry of Greenpoint. The district has a long history of attracting immigrant groups as witnessed here looking at the St. Stanislaus Kostka R.C. Church (Driggs Ave), founded on the large Greenpoint Polish community.

Plate 5.7 Williamsburg and Greenpoint: Field Trips 1979 and 1982. *Left* to *right*: Manhattan St, Greenpoint; St. Stanislaus Kostka Church (Driggs Ave), Greenpoint; two and three story brownstones (note the popular cast iron railings stoops); Williamsburg shops with Jewish 'feltly' hats (Sources: Raymond Rauscher; Brooklyn Historical Society; and City of New York 197-a Plan)

Further along is an example of the two and three story brownstones (note the popular cast iron railings and stoops) of Greenpoint. The most popular form of housing construction throughout New York in the 1800s and into the 1900s was brownstones. Today these buildings are highly sought after and thus fetch a commanding sale price. Beyond Greenpoint the renewing of Williamsburg (noted earlier in this chapter) has been in full swing for several decades. Here the shop with

Jewish 'feltly' hats in the Williamsburg shopping street is a reminder of the large Jewish population that for most of the life of Williamsburg has called this neighborhood home.

As noted earlier in this Chapter, in reviewing the Rheingold Brewery Plan, The Campaign for Community Based Planning (Municipal Arts Society) contributed to comments on the Williamsburg 197-a Plan program (within Brooklyn Community Board). These comments again were centered on the document *Planning for All New Yorkers: An Atlas of Community Based Plans in New York City Campaign* (Municipal Arts Society 2008). As noted earlier, this was the work of the community-based Planning Task Force (co-ordinated by the Municipal Art Society Planning Centre) (www.mas.org).

The Task Force, in an overview, noted there was a good deal of disinvestment in Williamsburg during the 1970s and 1980s. As a result population dropped and environmental health problems and poverty increased. The task force notes "artists started moving into Williamsburg in the mid-1980s and the neighborhood slowly started to revitalize."

The goals of the Williamsburg Waterfront197-a Plan aim to: waterfront development should conform to the scale and character of adjacent residential or mixed-use neighborhoods; increase waterfront access and public open space; promote local economic development; and, strengthen existing diversity to maintain a model mixed income, mixed-use community. Details of the recommendations can be sourced from www.home.nyc.gov.

The Task Force also offers several critical conclusions. One was that "none of the City's proposed parkland had been created (as recommended in the 197-a Plan). In addition most, if not all, of the privately owned sites have not been acquired." Another criticism was that the rezoning did not address any of the community's historic preservation goals. The report however notes that, since the rezoning, the Landmarks Preservation Commission had designated a number of new landmarks in Williamsburg/Greenpoint.

The Task Force had some positive notes too. It noted that the TrasGas Energy plan to construct the power plant at the Bushwick Inlet had been defeated. In respect to agency cooperation, the Task Force noted that the final plan modifications brought the plan into closer conformance with individual agency plans and policies, while respecting community concerns, goals and objectives.

The report continues in addressing the social and industrial situation in Williamsburg at the time under the *Williamsburg Waterfront 197-a Plan* (Department of City Planning, City of New York) (2013) www.home.nyc.gov "Today, Williamsburg has many distinct economically and ethnically diverse neighborhoods, each with different needs. The population has grown significantly since 1990, due to immigration and an influx of artists and young professionals. Although heavy industry has declined, the Williamsburg waterfront continues to be the site for new noxious uses, including waste transfer stations and the largest proposed power plant in New York city – 1,000 to 1,500 megawatts – at the Bushwick Inlet. Community opposition to these uses has been strong and historically successful."

Given these comments above and following the guidelines for 197-a Plan program (Chap. 4) it is useful to review how the plan making for Williamsburg and Greenpoint proceeded. At the request of Community Board 1 and local elected officials, the Department of City Planning proposed a zoning map and text amendments for an approximately 175 block area in the Greenpoint and Williamsburg neighborhoods within Community District 1 (Department of City Planning, City of New York 2013) (www.home.nyc.gov).

The City pointed out that this was separate from the Williamsburg Waterfront Rezoning of 2005 as the new plan aimed to protect the existing character of residential areas east of the 2005 rezoning area. The City noted the residential blocks that would be affected, as well as the boundaries of waterfront rezoning (2005). The City states that the rezoning aims to "preserve neighborhood character and scale by limiting the height of new development, to create opportunities and incentives for affordable housing through inclusionary zoning, and to support local retail corridors while protecting the residential character of nearby side streets". It is of interest that the blocks within the rezoning area were originally developed in the nineteenth and early twentieth centuries as residential neighborhoods to house workers attached to the vibrant industry located along the East River and Newtown Creek, the City points out. These industries included ship building, metal and glass production, and oil and sugar refining.

The City notes that industry in the area of Greenpoint and Williamsburg declined steadily throughout the 1900s. The area under planning consideration has seen considerable growth during the last decade as a residential neighborhood. In response to these changes, the City notes, industrial and mixed use areas on and near the waterfront were rezoned to permit residential development under the 2005 Greenpoint-Williamsburg Land Use and Waterfront Plan.

Today, the City points out, most blocks in the area as subject to this rezoning consist of 2–4 story wood-frame attached houses and apartment buildings, while some buildings rise to five or six stories. Neat rows of brick and masonry row houses can also be seen in Greenpoint east of Monsignor McGolrick Park is noted. Of special interest (the city notes) is the Greenpoint Historic District (designated in 1982), featuring distinctive nineteenth century brick row houses commissioned by business owners, foremen, and professionals who had worked on the nearby waterfront. Of typical Brooklyn layout, the City notes, these buildings often include ground floor commercial uses when located along the commercial corridors (these include the avenues of Manhattan, Driggs, Nassau, Graham and Metropolitan and the streets of Grand and Franklin). Its noted, the area, as in the past, has scattered industrial uses throughout (especially Northern Greenpoint and Metropolitan Avenue west of Bushwick Avenue).

On May 11, 2005, New York City's Department of City Planning approved a rezoning of 175 blocks in Greenpoint and Williamsburg. According to the project's Environmental Impact Statement, the rezoning is expected to bring approximately 16,700 new residents to the neighborhood by 2013. The City calculates this development would include: 7,300 new units of housing; 250,000 square feet (23,000 m^2) of new retail space as projected; a loss of just over 1,000,000 square feet (93,000 m^2)

of existing industrial capacity. The rezoning also included a 28-acre (110,000 m^2) waterfront park. Finally, the City notes, "included in the plan's requirements are provisions for a promenade along the East River, built piecemeal by the developers of existing waterfront lots."

An inclusionary housing plan was included in the City resolution, providing height bonuses along the waterfront, allowing up to seven-story height increases. Likewise, in Northside Williamsburg, developers who provide apartments at rates considered affordable for low-income households (below 80 % of the area's median income) would attract bonuses. The City rightly points out that the "rezoning was a dramatic change in scale to a previously low-slung, industrial neighborhood." It is interesting that, at the time the proposed changes were the subject of much debate, including a letter written by the late Jane Jacobs (noted in the 'acknowledgments' section of this book) to Michael Bloomberg, Mayor at the time, criticizing the proposed development.

The community's plan, the City points out, is generous by providing provisions for schools, daycare, recreational outdoor sports, and pleasant facilities for those things. The City further notes that "the community's plan does not promote new housing at the expense of both existing housing and imaginative and economical new shelter that residents can afford. The community's plan does not violate the existing scale of the community, nor does it insult the visual and economic advantages of neighborhoods that are precisely of the kind that demonstrably attract artists and other live-work craftsmen." The City however acknowledged that there has been concerns. These included a concern that the existing community's character would be changed as existing residents were forced to move, and, further, that public transportation and public safety infrastructure would be unable to accommodate the projected 40,000 new residents.

The City notes:

> A boom in construction followed the rezoning, leading to complaints from neighborhood residents and their elected representatives. The zoning plan was modified on March 2, 2006, to include anti-harassment provisions for tenants and add height limits in portions of upland Williamsburg. Neighborhood organizations made differing opinions known: the Greenpoint-Williamsburg Association for Parks and Planning expressed approval of the proposal (with reservations), but many neighborhood residents and members of Community Board 1 continue to voice their objections.

The Community Based Task Force (as noted above) also was responsible for commenting of the Greenpoint 197-a Plan. Within the document *Planning for All New Yorkers: An Atlas of Community-Based Plans in New York City Campaign* (Municipal Arts Society 2008) a number of observations on this planning program are made. One concerning observation is: From the 1950s to current times (early 2000s) "public policy has led to Greenpoint eastern sector becoming a dumping ground for noxious businesses and burdensome municipalities facilities. Along Newtown Creek alone there are over 500 acres of heavy industry, including waste transfer stations, a sewage treatment plant, the former municipal incinerator, the marine transfer station, and a range of facilities suspected of storing hazardous substances."

Observing the social circumstances the report notes: "There has been a dramatic rise in the demand for housing and related community facilities due to a steady stream of immigration in the last 20 years. Artists and young professionals have also been flocking to the neighborhood seeking large spaces and cheaper rents than can be found in Manhattan or the quickly gentrifying area of Williamsburg, just to the south."

The goals of the Greenpoint 197-a Plan aim to: increase and revitalize waterfront access and public open space; promote low-rise housing and commercial development while protecting Greenpoint's environment and quality of life; make the community more pedestrian and bicycle-friendly; maintain the historic character of the neighborhood. Details of the recommendations can be sourced from www.home.nyc.gov.

The Task Force, in reviewing the Greenpoint 197-a Plan offers several critical conclusions. Under rezoning the Task Force comments that there are serious concerns from within the community about the integrity of many of the City's intentions for the rezoning. The community sought, reported the Task Force, to lower the height and density of development allowed under the current proposal's zoning designations; to increase the percentage and mandate the affordability of housing developed; to better maximize waterfront access and open space opportunities, and to more strongly support viable clean, light industry and commercial businesses. On another note, the Task Force noted that there had not been any significant affordable housing away from the waterfront (at 2007). Finally, the rezoning did not protect existing manufacturing businesses outside the designated industrial business zone. The community wanted a 'mixed' use zoning to address the need, hence manufacturing buildings were being torn down "on a weekly basis", the Task Force commented.

Finally, the historical significance of the Greenpoint/Williamsburg rezoning planning area is born out in the work at the time of The Municipal Art Society. The Preservation committee and the community undertook a survey of the entire 184 blocks that the City proposed to rezone. Surveyors spent weekends walking the streets, photographing, convening discussions, and being guided by architectural historians. The exercise resulted in the MAS recommending 264 buildings eligible for listing on the State and National Register.

5.5 Urban Renewal Brooklyn Downtown

By 2013 the renewal of Bushwick was further advanced by the upgrading of Brooklyn Downtown (a transit ride by bus or train from Bushwick and other parts of Brooklyn). This downtown area holds a rich history and was always the focus of shopping and governance for Brooklyn. The scale of development plans by 2009 (reflected widespread residential, commercial and services interest in this part of Brooklyn (Plate 5.8), based partly on the *Downtown Brooklyn Development Plan* (City of New York 2004).

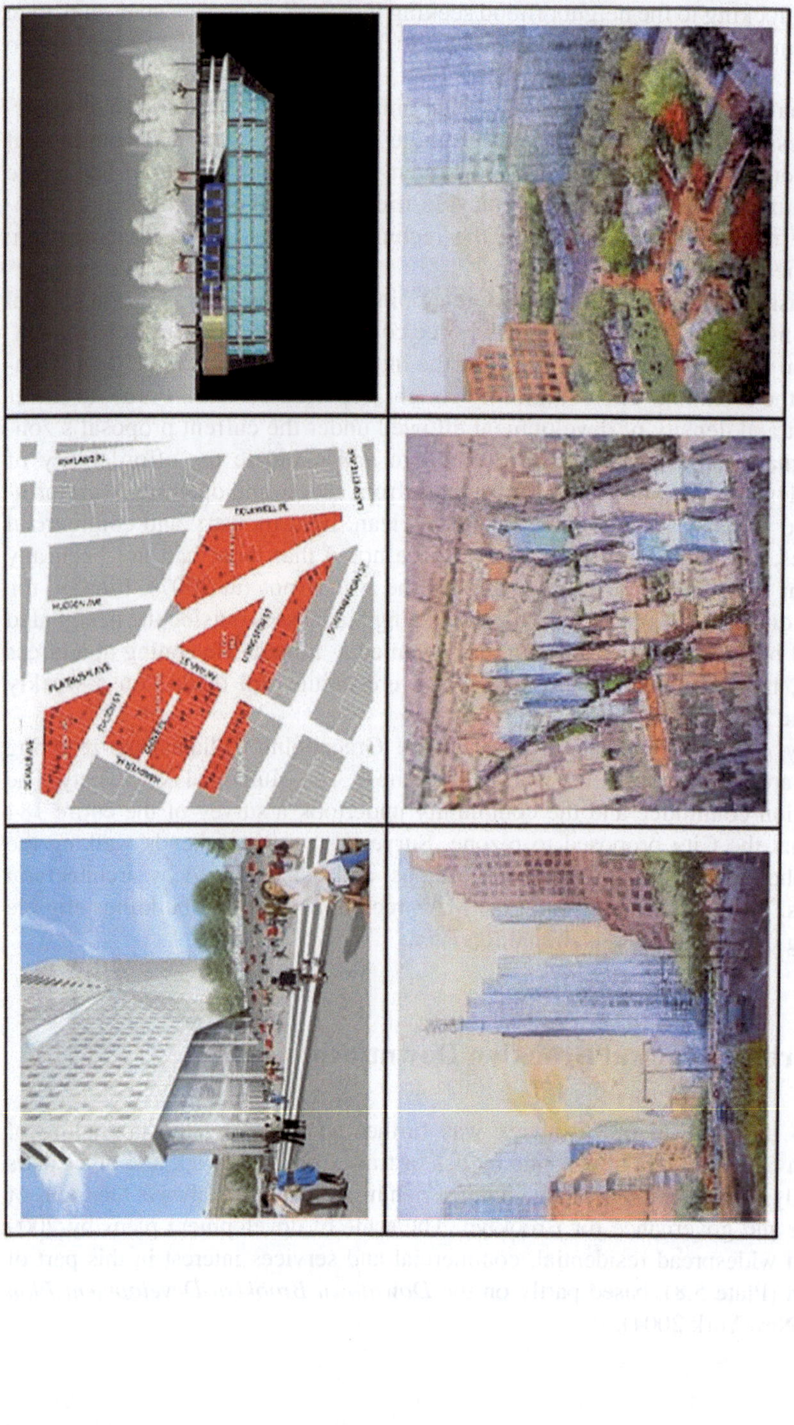

Plate 5.8 Downtown Brooklyn Proposals. *Left* to *Right*: BAM South Site Proposal; Eastern Gateway Residential; and Willoughby Park, Rendering of Underground Car Park; proposed residential high rise; layout of plans for renewal of Brooklyn Downtown; plan for Willoughby Park (Source: Downtown Brooklyn Partnership (2013))

The Downtown Brooklyn Partnership (DBP), a not-for-profit local development corporation, works closely with the City of New York administration, business and community interests. The DBP states on its website that it strives to make Downtown Brooklyn a world-class business, cultural, educational, residential, and retail destination. Examining the activities of the DBP gives an insight into the urban planning being progressed in the DBP (http://downtownbrooklyn.com).

The DBP states it is working together with the three business improvement districts (BID) that the DBP manages: the MetroTech BID; the Fulton Mall Improvement Association: and, the Court-Livingston-Schermerhorn BID. The DBP's includes in its diverse activities: attracting new businesses and improving the environment for existing companies; facilitating the construction of public spaces and streetscapes that promote an active and cohesive community; supporting and promoting Downtown Brooklyn's cultural assets; and encouraging a unified sense of place and an engaged civic community.

Thanks to significant public and private investments over the past decade, the DBP states, Downtown Brooklyn is now a key economic center for New York City. The Downtown development plans provide an overview of this expected investment (Plate 5.9).

The Downtown Brooklyn Partnership (DBP) notes the area boasts: a business district with more than 17 million square feet of office space; a historic and vibrant shopping center; a burgeoning residential community, 11 colleges and universities; nearly 50 cultural organizations; a professional sports franchise; and, exceptional public transportation and open spaces.

To further complement the more than $300 million that has been invested in Downtown Brooklyn by the City of New York, the DBP works with local stakeholders to advocate for streetscape improvements across the area. In 2012, the DBP unveiled a strategic plan to maximize the area's economic potential and further the growth of Downtown Brooklyn. At the event, DBP notes, the New York State Comptroller Thomas DiNapoli released a report stating the job growth of Downtown Brooklyn, where no jobs were lost during the recession and employment actually grew 18.3 % between 2003 and 2010.

Later that year the DBP outlined a strategic plan for Downtown Brooklyn that focuses on five key priorities: Connectivity, Diversification, Community, Culture, and Opportunity. The DBP also celebrated Downtown Brooklyn's position as America's fastest-growing college town and residential neighborhood, home to nearly 50 renowned cultural groups, and a thriving diversification of businesses.

Downtown Brooklyn, the DBP notes, is home to nearly 50 world-class arts and cultural organizations that play a central role in the community and economic development of the area. While the highly-acclaimed Brooklyn Academy of Music (BAM) is one of the best-known cultural venues located in the area, scores of other organizations add to Downtown Brooklyn's unique character and quality of life.

The DBP keeps a master plan for the Cultural District, a vibrant, multicultural arts district in the neighborhood surrounding the BAM. This effort involves the conversion of underutilized, city-owned properties into affordable performance and rehearsal space for a diverse array of non-profit visual, performing, and media arts

Plate 5.9 Closeup of Downtown Brooklyn Development Plans (Source: Downtown Brooklyn Partnership 2010)

groups. The DBP advocates for infrastructure improvements and strategic long-term planning, and raises the visibility of the area's performance facilities and cultural institutions.

The first phase of the District's development, DBP notes, involved the renovation of the 80 Arts – James E. Davis Arts Building, which was completed in Summer 2004, becoming the Cultural District's first completed project. The 30,000-square-foot building is home to 12 diverse nonprofit arts groups benefiting from below-market rents and shared amenities.

In December 2012, then Mayor Bloomberg announced three major milestones to further strengthen the cultural community in Downtown Brooklyn. These three development projects included:

1. BAM North Site 1: Site to be developed by developers in partnership with the City Department of Housing Preservation and Development (HPD). The 515,000- square-foot mixed-use development will include 600 residential units, of which 50 % will be affordable, 20,000-square-feet of cultural and related office space, and 20,000 square feet of retail space.
2. BAM North Site 2: Then Mayor Bloomberg and HPD announced in October 2013 the selection of a team to develop an affordable, mixed-use development which will create 42 units of affordable housing along with new ground-floor commercial space for a restaurant and 27,000-square-feet dedicated to a cultural space to be designed by Eyebeam and Science Gallery.
3. BAM South Site: In 2013, a project to develop the site was approved by the New York City Council. The 32-story mixed use cultural development will include 60 affordable housing units and 240 market rate units, over 15,000 square feet of retail, 50,000 square feet of cultural space, and a new public plaza. The cultural space, which is a critical component to this project, will be allocated to a BAM cinema, rehearsal space for 651 Arts, and a new library.

In November 2012, the DBP and its partners at the DUMBO Improvement District and Brooklyn Navy Yard designated a consortium to develop the Brooklyn Tech Triangle, an area home to more than 500 tech and creative companies employing 9,628 people – all of which is poised to double over the next 3 years. In order to ensure growth in the Tech Triangle is strategic, the Strategic Plan – unveiled in 2013 – addresses issues related to place making, land use, transportation and infrastructure, and real estate and economic policy.

The investments made in Downtown Brooklyn 2006–2013 (Plate 5.9) were: $5.2b in private investment; 8,060 residential units; 1,570 hotel rooms; 1.5 m square feet of retail space; 432,000 square feet of commercial space; and, 12.8 million square feet of development.

In recent years, Downtown Brooklyn has attracted major public and private investments, including Barclays Center, the Cultural District, and NYU's Center for Urban Science and Progress. Yet, as DBP notes: "the Eastern Gateway (Plate 5.8) which connects these assets contains a number of buildings where the ground and upper floors have long been vacant. Bringing these spaces back online will help facilitate economic growth in this area and build a cohesive Downtown Brooklyn community. To help incentivize this transformation, the Downtown Brooklyn Partnership received a grant from the New York State Division of Housing and Community Renewal that will leverage private sector investment to renovate vacant spaces within the Eastern Gateway and catalyze economic growth and job creation in the area."

The proposed Willoughby Square is the long-awaited public space that New York City officials consider Brooklyn's answer to Bryant Park in Midtown Manhattan. An entirely invisible world underneath the park (Plate 5.8) in the form of a high-tech, underground garage, DBP notes, will in part finance creation of the above ground square, an agreement negotiated by the city that resurrects a cornerstone of the extensive plans for rezoning the borough's downtown.

The DBP notes the square will span more than an acre on Willoughby Street, a half-block from the Fulton Street mall, with manicured lawns, walkways and gardens, as well as a site to commemorate the abolitionist movement. Below it, the garage will house about 700 cars at any given time, relying on a system of computers rather than garage attendants. It will be the largest automated parking facility in North America.

5.6 South East Brooklyn (Including Coney Island) – Urban Planning Directions

The City of New York is also completing plans for the revitalization of areas such as Coney Island, a major amusement area about ½ hour drive from Bushwick. The Coney Island part of Brooklyn always drew millions of locals every summer until the area became blighted in the 1970s and 1980s. The City adopted a comprehensive plan to encourage new development and guide the future growth of Coney Island in 2005 (Plate 5.10). By 2013 the city was looking at encouraging up to 5,000 new housing units outside the immediate amusement area. As with Brooklyn downtown revival improvements to Coney Island would economically benefit all Brooklyn, including Bushwick.

Some background notes on this planning can be summarized (Department of City Planning, City of New York 2013) (www.home.nyc.gov).

The City notes: "From its famed boardwalk to the iconic Wonder Wheel, Cyclone, and Parachute Jump, to the once bustling midways along Surf Avenue and the Bowery, Coney Island's legendary status as the world's greatest urban amusement park lives in all of our memories. But over the years, this once vibrant area has shrunk and deteriorated, and is limited today to one block of seasonal amusements and a few remaining active frontages along Surf Avenue."

The Department of City Planning, in partnership with the New York City Economic Development Corporation (EDC), the Department of Parks and Recreation (DPR) and the Department of Housing Preservation and Development (HPD), has developed a comprehensive plan that will establish a framework for the revitalization of the Coney Island amusement area and the surrounding blocks, within the *Coney Island Comprehensive Rezoning Plan* (City of New York 2009). The plan builds upon the few remaining amusements to create a 27-acre amusement and entertainment district that will re-establish Coney Island as a year-round, open

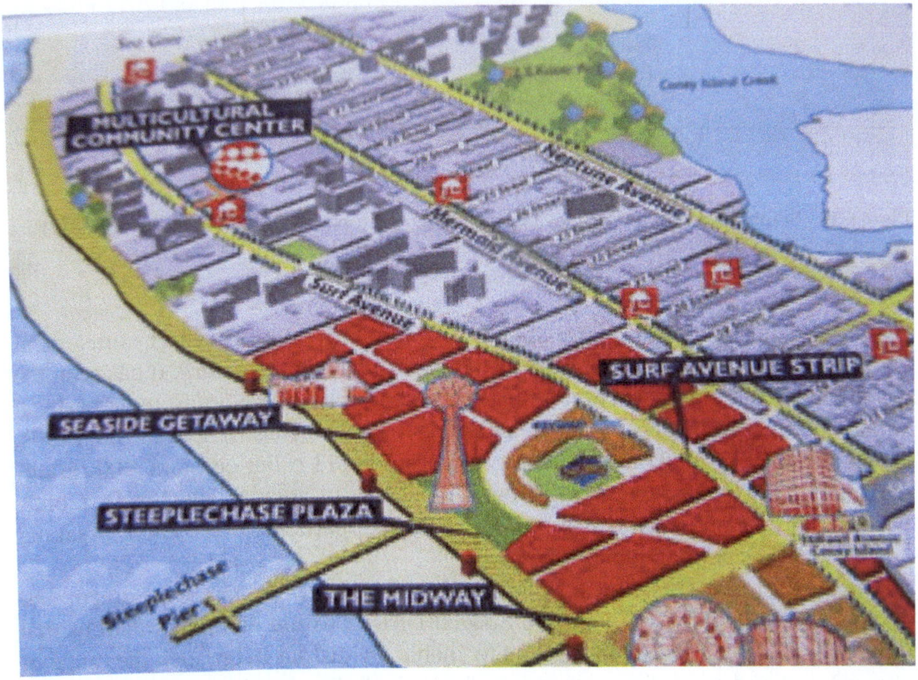

Plate 5.10 Southeast Brooklyn (Coney Island) Plan (Source: City of New York 2013)

and accessible amusement destination. Outside of the amusement area, the plan provides new housing opportunities, including affordable housing, and neighborhood services. The plan covers 19 blocks bounded by the New York Aquarium to the east, West 24th Street to the west, Mermaid Avenue to the north and the Riegelmann Boardwalk to the south. The plan seeks to, according to the City:

1. Facilitate the development of a vibrant year round, 27-acre urban amusement and entertainment district by catalyzing a variety of new indoor and outdoor amusement, entertainment, and complimentary uses, and laying the groundwork for the development of a 12-acre urban amusement park preserving and expanding amusement uses in their historic boardwalk location in perpetuity;
2. To the north and west of the amusement area, foster the redevelopment of vacant and underutilized land, providing opportunities for new and affordable housing as well as a broad range of neighborhood retail and services that the Coney Island community has lacked for decades;
3. Create a vibrant pedestrian environment, with Surf Avenue serving as the reinvigorated retail and entertainment spine of the district;
4. Recognize and support the Coney Island's unique character, culture and needs through the creation of the Special Coney Island District;
5. Through the development of year-round uses and job opportunities for the residents, facilitate the economic revitalization of the peninsula.

Plate 5.11 Southeast Brooklyn (Coney Island) Planning Proposals. *Left* to *Right*: Wonder Wheel Way; Reactivated Bowery; Coney East; and, Mermaid Ave (Source: City of New York 2013)

Coney Island's emergence as a world renowned, one-of-a-kind amusement destination, the City states, dates back to the mid-nineteenth century. It has seen the development and the destruction of that era's most well-known amusement parks in America, including Luna Park, Dreamland and Steeplechase Park. In the 1930s, Coney Island contained 60 bathhouses, 13 carousels, 11 roller coasters, 200 restaurants and 500 businesses ranging from newsstands to arcades and hotels. Coney Island's spirit is one of accessibility: everyone can participate or watch, breaking down social, gender and racial barriers.

Coney East, the City notes, is intended to become a 27-acre amusement and entertainment district. The proposed Special District would define regulations that modify those of the underlying zoning to enable the district to broaden the range of allowed uses and create four categories of uses, including hotels; open and enclosed amusements with retail; restaurants of all sizes; and retail and service uses.

To promote a vibrant active district, the City states, and to create a seamless transition between the amusement park and the rest of the entertainment district, new developments are proposed within the planning (Plate 5.11), including: Wonder Wheel Way (a proposed new street); reactivated Bowery; Coney East; and, Mermaid Ave. The planning envisages amusements occupying half of the total street frontage and the ground floor level of hotels having active uses such as restaurants, retail and entertainment venues.

Coney West will be rezoned, according to the City, to allow for residential uses with ground-floor retail along Surf Avenue and revitalized uses along the Boardwalk. On Surf Avenue, commercial ground floor uses would be mandated and 20 % of the frontages would have to be dedicated to entertainment-related uses. Along Surf Avenue Special, special district regulations will mandate that buildings locate at the street line. The minimum and maximum base heights for Surf Avenue will be 60 and 85 ft, which is the height of the Shore Theater. Towers would be allowed in defined locations, with the highest tower placed at the corner of Surf Avenue and the intersecting streets. To preserve the unique sense of openness on the Boardwalk, special regulations will limit uses to amusement and entertainment and cap building height at 40 ft, which is the height of the landmarked Childs' building.

Since the closing of Steeplechase Park in 1964, The City points out, the amusement area has significantly shrunk, consisting today of only a few blocks of largely seasonal amusement attractions. Today, the City states, and since the closing of Astroland at the end of the summer 2008, the amusement area consists of one block of largely seasonal amusement attractions and a few active frontages along Surf Avenue. Deno's Wonder Wheel Park remains the only open amusement park in Coney Island at that time.

Some of the historic amusement structures remain and lend an iconic presence in the largely vacant rezoning area, the City acknowledges. A number of these structures are New York City landmarks: the Cyclone, Wonder Wheel, Parachute Jump and the former Childs restaurant. Despite its decline, Coney Island's amusement area continues to attract thousands of visitors every year, the City states, demonstrating the areas potential and its unique legacy as an urban beachfront amusement destination. The seaside location, easy access to public transit, and extensive Boardwalk remain its unique assets.

Key Span Park occupies the central portion of the rezoning area and was built by the City of New York in 2001 as the home of the Brooklyn Cyclones, a New York Mets minor league baseball team. The City notes that Key Span Park attracts thousands of visitors a year during the summer baseball season, which runs from the end of June through early September. To the west lie two parking lots, totalling approximately 405,000 square feet in area and currently mapped as parkland, that serve the Brooklyn Cyclones during baseball season and are dormant the rest of the year. The Abe Stark Skating Rink, which serves ice hockey leagues, is located on the boardwalk frontage of one of these parking lots. An underused Green Thumb garden fronting the Boardwalk shares the southern end of the other block.

The comprehensive plan for Coney Island, the City acknowledges, builds upon the commitment made by Mayor Bloomberg in 2005 with the announcement of the Coney Island Strategic Plan. The new plan, the City states, is the product of over 300 public meetings with numerous stakeholders, ranging from elected officials, residents, property and business owners, Coney Island enthusiasts from New York City and beyond. The City's plan will help to ensure that future generations can enjoy an open, affordable, urban twenty-first century Coney Island that does justice to its illustrious history and enduring appeal.

5.7 Summary and Conclusions

This chapter looked at the application of New York City planning instruments, including the 197-a Plan process. How these planning instruments were applied to Bushwick, Williamsburg, Greenpoint, Downtown Brooklyn and South East Brooklyn were examined as case studies. Differences in the applications were explained, such as the Bushwick planning focus centering on the Brownfield Redevelopment Plan program. The chapter reviewed this program as applied to the West Bushwick area (Rheingold Renewal Project). Next, the 197-a Plan applications to Williamsburg Waterfront, Greenpoint and combined Williamsburg/Greenpoint area were examined (including historical backgrounds). Finally, to examine other planning instruments, such as master planning, two other areas (Brooklyn Downtown and South East Brooklyn, including Coney Island) were examined.

References

Brooklyn Historical Society Archives (2005)
Bushwick Observer, Newspaper (2000)
City of New York (2000) Greenpoint-Williamsburg Contextual Rezoning. City of New York, New York City
City of New York (2002a) Greenpoint 197-a Plan. City of New York, New York City
City of New York (2002b) Williamsburg Waterfront 197-a Plan. City of New York, New York
City of New York (2004) Downtown Brooklyn Development Plan. City of New York, New York City
City of New York (2009) Coney Island Comprehensive Rezoning Plan. City of New York, New York City
City of New York Archives (2013)
Downtown Brooklyn Partnership Archives (2013)
Municipal Arts Society (2008) Planning for all New Yorkers: an Atlas of Community Based Plans in New York City Campaign. Municipal Arts Society, New York.

Chapter 6
Education: Preparing Students for Urban Futures

6.1 Introduction

This chapter looks at the evolution of high school education to prepare students for engagement in urban changes. Bushwick has 33 public and private, primary and secondary schools. This includes 14 public elementary schools, one charter school, four parochial schools, seven high schools, and one secondary school. The chapter focuses on how schools are preparing students for a workforce that will face urban planning challenges. The chapter starts by reflecting back on the Class of 1961 at Bushwick Highschool. The creation of a New Century School (Academy of Urban Planning) on the Bushwick High School campus is examined. Two other special schools located in Bushwick are examined, Bushwick Leaders' High School for Academic Excellence and the Brooklyn Latin School.

6.2 Bushwick Schools

Good schooling in America has always been considered a right of every person living in America. The commitment to education in Bushwick is demonstrated by the school buildings themselves. A selection of schools in Bushwick, including those to be examined more closely, is contained in Plates 6.1 and 6.2. Note the general grand style of the buildings and the variety of architecture. Those schools contained in Plate 6.1 include: Public School 86; Public School 123; Bushwick High School Campus; Cabrini Catholic School (Suydam St); The Brooklyn Latin School (TBLS), TBLS insignia.

The Bushwick schools and subjects in Plate 6.2 include (left to right): EBC Highschool for Public Service; Saint Elizabeth Seton School; newly located The Brooklyn Latin School; Academy of Urban Planning AUP students Class 2007; ACP staff member Adam Schwartz (center) with Meryl Meisler (artist/photographer) and Raymond Rauscher (on field trip) holding the Up From Flames cover; and,

R.C. Rauscher and S. Momtaz, *Brooklyn's Bushwick - Urban Renewal in New York, USA: Community, Planning and Sustainable Environments*, DOI 10.1007/978-3-319-05762-0_6,

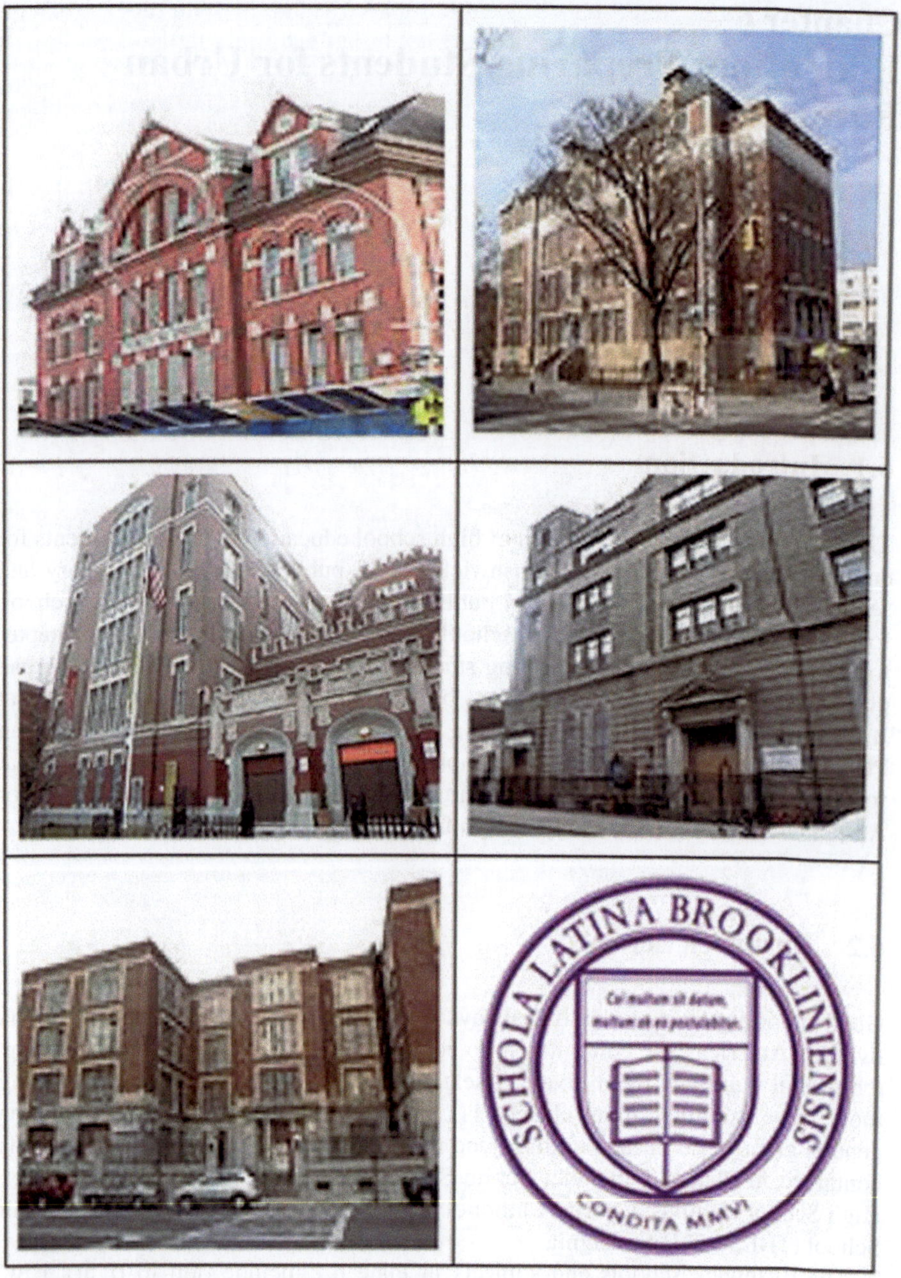

Plate 6.1 Bushwick Schools Part 1. *Left* to *Right*: Public School 86; Public School 123; Bushwick High School Campus; Cabrini Catholic School (Suydam St); The Brooklyn Latin School (TBLS) 436 Bushwick Ave; Insignia of TBLS (Source: wiki-media (13 Nov 2013))

Plate 6.2 Bushwick Schools Part 2. *Left* to *Right*: EBC Highschool for Public Service; St. Elizabeth Setton; new Latin School location, Bushwick; Planning AUP students Class 2007; ACP staff member Adam Schwartz (*center*) with Meryl Meisler (artist/photographer) and Raymond Rauscher (on field trip) holding the Up From Flames cover (2008); Rheingold Renewal Site international design team and local participants at workshop (attracting AUP interest) (Sources: wikimedia (13 Nov 2013), AUP Class Book 2007, and Raymond Rauscher)

Rheingold Renewal Site international design team and local participants at workshop (attracting interest of all high schools of Bushwick).

6.3 Bushwick Class of 1961

Bushwick High School has always been the central institution of high school education in Bushwick. The *Yearbook of Bushwick HS 1961* is a reference that tells part of a story of life in school at that time. The Principal and the department heads at that time reflected predominately older and mostly male teachers. Teaching embraced core subjects of English, social studies and science. Teachers dress was predominantly suit and tie for males and dresses for females. Most department heads had their own rooms and department responsibilities, including curricular development (still meeting Regents Exam regulations).

Some insights into living in Bushwick in the 1940s–1960s is presented below from seven participants (names left out for privacy) from the Class of 1961 who during the book's research shared stories with the author (photos of participants and the school principal in 1961 are in Appendix 1 Plate A1).

Participant 1

"We grew up in Williamsburg, a beautiful section, but our street on Bushwick Ave (and Arion Pl), was surrounded by poverty. We lived in an apartment house with no heat. Our parents remembered the lean hard years of the depression, so they refused to move to a better area. Dekalb Ave, about a mile walk from our home, had beautiful homes and landscapes. Our family did not have a car so we travelled everywhere by subway, bus or walked. We dragged all our groceries in a shopping cart, thus increasing our stamina. We felt safe in our own area, taking trains and buses and coming home late without feeling afraid."

Participant 2

"A lot of my memories about childhood are not clear as I would like. I certainly played lots of games. My aunts taught me Canasta and we played for days at a time. I loved to roller skate and jump rope. I played baseball at PS 57 (Troutman St) during the summer programs which the city ran. We played high water, low water, leaping over rope pulled high and higher. The boys did stickball in the street and built scooters with wheels from old skates. We learned about sportsmanship, compromise, fair and unfair play, rule making and friendship. The only parental rule imposed was how far one could go in distance and how late could one stay out. We were allowed to be kids in a way often today denied to children."

Participant 3

"I loved Fulton St and Knickerbocker Av for shopping. There were no credit cards; shopping bags were brown paper with twine handles. Every household had a shopping cart for weekly shopping. You shopped by foot, maybe bus, but only third by car (if families had one)."

Participant 4

"As I turn pages of our 1961 yearbook I see kids from humble backgrounds. I am nostalgic for the civility of those days. Our teachers held a genuine care for the students, as I'm sure they do today. Here were no overhead projectors, visual aids of any kind, no computers. These teachers developed in me a love of knowledge and sound work habits, helping me move out of grimy surroundings to a new life."

Participant 5

"In 1950 our street, Cedar St, had not changed in decades. The stick ball games would fill the summer air, day and night. Our homes were a solid array of timber reflecting a working class. Many families were renting, as we did. I can still hear the call from mothers looking for kids to go to the corner shop for daily needs. The Good Humor ice cream man turned the corner like clockwork every summer day, as did the watermelon man and knife sharpener man."

Participant 6

"St Barbara's church remains one of the buildings I loved. Most of the others I knew so well are gone. I have never been able to find Italian ice like they had at Stella Doros, nor have I been able to buy soft pretzels out of a baby carriage. I remember the pictures of the Rheingold beauty contestants landing in the grocery stores. It was here where you could buy bologna on a hard roll with real German potato salad. Food brings nostalgia, but its Rock and Roll that really holds special memories. 'Sixteen Candles', 'In the Still of the Night' and the local doo-wop harmony could be heard on the street corners."

Participant 7

I lived only two blocks from Bushwick High School and could see the building from my private 'tar beach' in the summer. It was somehow always comforting to see the quasi medieval style building in the distance. It represented stability, strength, and comfort qualities. As a teenager these were important to me. Many of my teachers in junior high and high school were politically liberal. I believe their ideas and values played a role in the development of my political leanings.

6.4 New Century School (Academy of Urban Planning)

The change in Bushwick High School's socio-economic group paralleled the changes taking place in Bushwick in the 1960s and 1970s. By 2000 an educational experiment was underway aimed at breaking down big and under-achieving schools to smaller and more personable ones. This program, with the support of Bill Gates

Foundation, would see a number of New York City schools broken down into smaller schools. Bushwick HS is thus today three schools in one: Academy of Urban Planning; School of the Environment; and, School of Social Justice. An earlier Harbor School initiative was successful and moved to a more suitable location closer to water.

The New Century School of the Academy of Urban Planning opened its doors in 2003, with the student population at about 500 by 2007. In the 2008 graduating class 85 % would continue their education (2/3rds to 4 year colleges or universities and 1/3rd to 2 year colleges or technical schools).

The Academy, different from schools in the past, provides students with experience through partnerships with organizations throughout the city. These partners work in the classrooms and create internships. The partners also advise on curriculum development and offer resources for projects and activities. In 2007 the partners included (AUP Sourcebook 2007):

The difference of the Academy to other schools is also in the curriculum. Students here, for example, attend National Planners Network Conferences. The emphasis here is to bring youth into the decision making processes at municipal and regional levels, especially where schools and communities are involved. The courses offered at the Academy usually emphasize practical learning experiences, for example (AUP Sourcebook 2007):

Background on work completed by students at the Academy of Urban Planning is contained in Plate 6.3. These works include (left to right): a community mural; an art and urban design; a geographic information systems; an urban ecology project; a history project; and, the Planners Network Conference.

6.5 Other Innovative Bushwick Highschools

There are a number of high schools located in Bushwick that are experimenting with new teaching methods. The Bushwick School of Social Justice (BSSJ 2013), a second new century public high school, is one of four schools currently occupying the Bushwick Campus. Enrolment at the BSSJ is approximately 425 students (2012). The school is partnered with: Make the Road New York, Brooklyn College, and the Institute for Student Achievement (ISA). It was founded by Terry C. Byam, Matt Corallo, Lorraine Gutierrez, Matthew Ritter, and Mark Rush. It opened its doors in 2003, graduated its first class in 2007, and has received an ‘A’ rating for the last 5 years. Terry C. Byam was the founding principal and the current principal is Mark Rush.

The Bushwick Leaders’ Highschool for Academic Excellence (BLHAE) is a specialized high school preparing students to become active leaders and was founded in 2003. The school web site provides the following details of the school:

> East Brooklyn Congregations (EBC) is an example of congregation-based community organizing serving several neighborhoods in New York City. Formed in 1980, it is affiliated

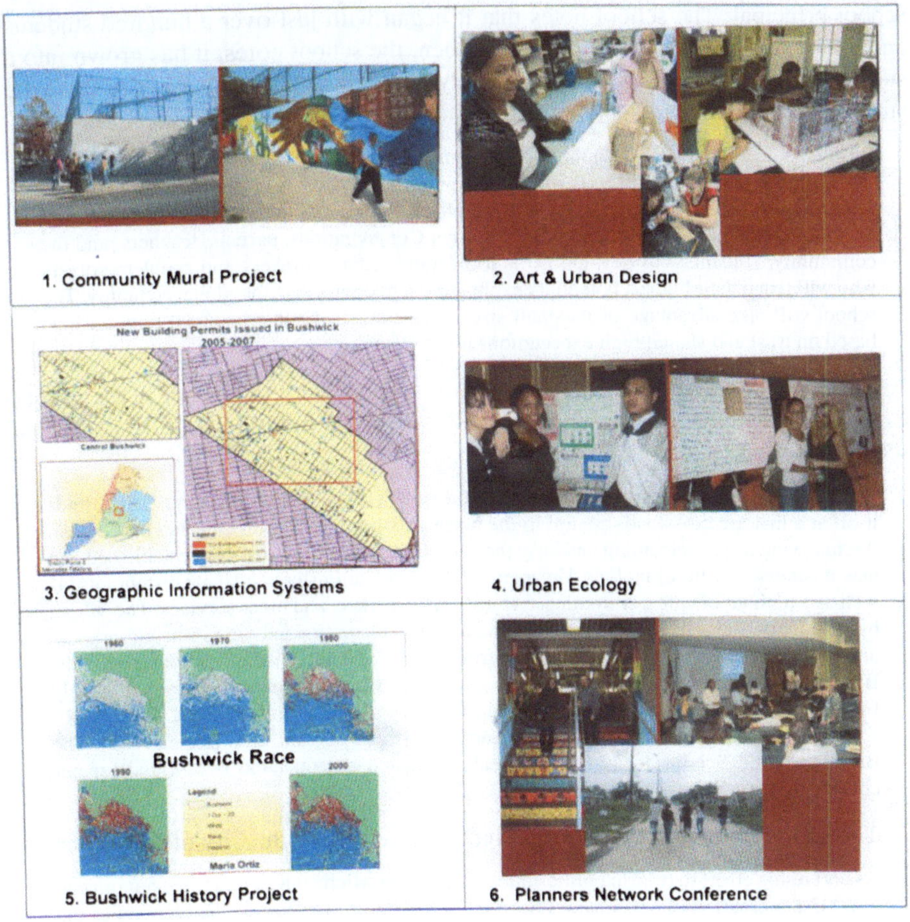

Plate 6.3 Academy of Urban Planning Student Projects. *Left* to *Right*: a community mural; an art and urban design; a geographic information systems; an urban ecology project; a history project; and, the Planners Network Conference (Source: Academy of Urban Planning Source Book 2007)

with the Industrial Areas Foundation and headed by Michael Gecan. EBC is best known for founding Nehemiah Homes and building 2,100 houses that low-income families could afford to buy. It has been involved in the formation of three public New York City high schools: EBC East New York, EBC Bushwick High School for Public Service, and Bushwick Leaders' High School for Academic Excellence.

BLHAE first opened its doors in the fall of 2003. It was the result of collaboration between the New York City Department of Education, East Brooklyn Congregations (a community-based organization that specializes in housing, public works, and educational projects). Leadership was offered also by Catherine Reilly, school principal. The school notes that it began with just over a hundred students and a bare, cinder-block building. Since then, the school notes, it has grown into a successful high school in a historic Brooklyn monument that endeavors to serve the students in the Bushwick community.

> Bushwick Leaders' High School for Academic Excellence will prepare ninth through twelfth grade students from the Bushwick Community, regardless of prior academic performance, to become active leaders with an emphasis on academics, college preparation and technology. In collaboration with East Brooklyn Congregations, parents, teachers, and the community, students will develop skills to become critical thinkers and problem solvers who will reach their highest potential enabling them to compete in the global economy. The school will take advantage of its small size to develop strong interpersonal relationships based on trust and shared high expectations for students.

The Brooklyn Latin School (TBLS) is a specialized high school in New York City, founded in 2006. The school web site provides the following details of the school as follows.

> The School is situated on the border between Bushwick and Williamsburg, TBLS finds itself at a historic crossroads central to the borough's history. TBLS is located on Graham Avenue, which is a welcoming and busy thoroughfare, with a diverse mix of established and new businesses. (These) include Mexican, Italian, Middle Eastern and Thai restaurants, as well as a number of cafes, convenience stores, supermarkets and other services. The school has routinely found the neighborhood to be a safe, welcoming and vibrant neighborhood, and it is easily accessible by the L train from Union Square, several bus routes, and the Brooklyn-Queens Expressway. The ideals governing Brooklyn Latin are borrowed largely from the Boston Latin School, and popular society's Ideals. John Elwell, the school's founder, and Jason Griffiths, administer and monitor the school. Like Boston Latin, The Brooklyn Latin School began with the realization that an education is essential for our country to retain its role as a world leader.

The TBLS continues, providing insight into the role of the City of New York:

> As part of his effort to reform public education, Mayor Michael R. Bloomberg announced in 2005 the creation of seven new selective high schools. To oversee the creation of the first of these new schools, Mayor Bloomberg selected the President of Replications, Inc., a not-for-profit organization that replicates successful school models around the country. In the fall of 2006, after arduous months of preparation, The Brooklyn Latin School opened its doors to 63 students who represented every borough of New York City, and countries as diverse as Poland, Nigeria and Israel. Moreover, Griffiths had recruited an energetic and capable staff whose experience in education, the private sector and public sphere had prepared them for the challenges of starting a new school. The Brooklyn Latin School is a thriving educational community, home to hundreds of students who on a daily basis validate the vision of its founders. And every February, the students of The Brooklyn Latin School brave the blistering New England cold and venture to Avenue Louis Pasteur in Boston. Sitting in the Boston Latin auditorium, surrounded by portraits of the great men and women who have walked its hallways, they are reminded that though their own school may be new, they are in fact part of a tradition that is nearly 400 years old.

Perhaps a poem (Plate 6.4) could best sum up the quiet revolution of educational movements, summarized in this Chapter examining Bushwick. Seeds are often planted many years previous, by school principals and teachers who pass on the ideals of academy and hope for all generations.

What makes a saint, what value knowledge imparted, what measures greatness
Clara Molendyke, Principal of Bushwick Highschool, 1950s/1960s working class times
A special era, New Yorkers responding to '40s war, reigniting renewed education thirst
Homes were happy again, feet hitting the streets of bounce, though budgets no frills

The teachers, Selz, Ellison, Abraham, Regal, poured studies out as smooth scotch
Time of new hope, no security checks, no truancy, no drugs, minds as open portals
Clara and team instilled inquiry, fairness, fellowship, national pride and justice all
May that lasting spirit of Bushwick infuse today's precious generation and more

Plate 6.4 Quiet Education Revolutions (Source: Raymond Rauscher 2013 (unpublished))

6.6 Summary and Conclusions

This chapter looked at the evolution of high school education at Bushwick, focusing on the changes to meet the challenges of urban futures. Firstly, the evolution of education at Bushwick High School 1950s/1960s was examined. Here the Class of Bushwick High School 1961 expressed their views on living in Bushwick and the value of education, then and now. The creation of the experimental New Century Schools in 2000 led to a look at the Academy of Urban Planning and School of Social Justice on the Bushwick Campus. The authors also examined other education school experiments in Bushwick, including Bushwick Leaders' Highschool for Academic Excellence (BLHAE 2013) and The Brooklyn Latin School (2013). The chapter ended on a note of looking at new advances in education to meet student needs in facing urban futures in any city, town or region. Chapter 7 looks at lessons learned from Bushwick and Northeast Brooklyn's urban renewal and future directions of embracing sustainable urban planning (SUP).

References

Academy of Urban Planning. www.aupnyc.org. Accessed 6 June 2013
Bushwick Leaders Highschool for Academic Excellence. www.bushwickleaders.org. Accessed 25 July 2013
School of Social Justice. www.newyorkschools.com. Accessed 11 July 2013
The Brooklyn Latin School. www.brooklynlatin.org. Accessed 22 July 2013

Chapter 7
Future Directions: Sustainable Urban Planning (SUP)

7.1 Introduction

This Chapter sets out the lessons learned from the history of New York City urban planning, centered on Bushwick and Northeast Brooklyn. The lessons focus on the engagement of communities and businesses in urban planning, including effectiveness of community boards and planning instruments. In particular, the future planning directions of Bushwick and Northeast Brooklyn are addressed. Finally, comments are offered on the challenge before governments to adapt their urban planning approaches to current times. Emphasis here is on challenging all governments anywhere to engage the community (including schools) and business sectors to consider a sustainable urban planning (SUP) approach.

7.2 New York City Planning Challenges

The urban planning challenges of a City as big as New York can be overwhelming. The City shows it is aware of the importance of community decision making (i.e. community boards) and businesses engagement (i.e. business improvement districts) (Chap. 4). This understanding has often been translated into governance legislation and subsequent planning processes (such as the array of planning instruments reviewed in Chap. 4).

The long history of establishment of the community boards in New York City is an example of the perseverance of those with a vision (in all sectors: community, business and government). The governance support of those boards in the planning of their districts (often places highly impacted over decades) is essential. In reviewing the establishment of these boards, and how they operate, it was apparent that they are essential to bringing 'city hall' to the neighborhoods. The administration structures that would be needed to ensure the boards were serviced and capable of long life existence were evident in the framework the City of New York established, however updating is critical.

R.C. Rauscher and S. Momtaz, *Brooklyn's Bushwick - Urban Renewal in New York, USA: Community, Planning and Sustainable Environments*, DOI 10.1007/978-3-319-05762-0_7,

The community boards, viewed within their planning work in Brooklyn, appeared to be inclusive of the population and progressive in the urban planning they were engaged in (Chap. 5). The opportunity for residents and business people to be directly involved in decisions affecting their neighborhoods appeared to have significant success in the urban planning applications examined (i.e. brownfield redevelopment and 197-a Plan applications). The selection by the City of planning instruments appeared to be ever wider and covering greater flexibility. Chapter 5, where the application of several of the key instruments was reviewed, showed the complexities of implementation and complications of reaching consensus in best outcomes.

The environmental impacts on an area (sometimes up to a century long) reflected neglect by City of New York and New York State governance over a long period. The engagement of community boards (including the business sector's participation) is concluded as being a sound means of focusing on a district's environmental concerns and reaching actions to address those concerns. The environmental impact concentration of waterfronts such as in Greenpoint (i.e. Newtown Creek) and Williamsburg (i.e. old industrial sugar refineries) presented (and still presents) challenges to government and the community. The examination of 'green design' approaches and creation of new environment protection bodies (incorporating alternative energy approaches) were being increasingly considered. Both the City of New York and the State of New York indicated steps were being taken to address a range of environmental challenges. The guidelines for taking a sustainable urban planning (SUP) approach to these questions (and more) were outlined in Chap. 3.

The critiques of the planning instruments, in particular brownfield redevelopment and 197-a Plan applications, showed weaknesses as well as innovative aspects. The urban planners' and community advocates' writings on the subject (Chaps. 4 and 5) provided a significant breadth and depth of review. One common criticism the writers held was the time delay (under elected officials and authorities) to sum up a need and to act on addressing that need. The dramatic example of urban decline in a number of Brooklyn neighborhoods as examined and the failure to act here was picked up in unison by the writers (as quoted in the chapter). On the other hand, so too were examples of how swift those same governments (city, state and federal) acted when woken to the need to act under dramatic circumstances.

When City of New York planning instruments were applied, their application in general responded well to the needs of neighborhoods on the ground. It was obvious that in most applications of programs it was the housing needs (affordable) that were a high priority (after community safety and health concerns). It was apparent there was a wide variation in the planning instruments available, often making the decision on what instrument to apply to an area a challenge to the City administration. One example of inaction was summed up by a writer who stated that brownfield sites were not redeveloped because rehabilitation costs were said to be too high. Thus, he wrote, developers shied away and City administrations claimed the budget could not afford the works. Finally, it is concluded complexity of the urban and environmental challenges facing districts and the City of New York were widely disparate, resulting in many 'long suffering communities' neglected over time.

When cities move into decline (as witnessed in New York in the 1960s and 1970s) the administrations at all levels of governance (city, state and federal) need to come up with measures to reverse that decline. The situation in New York, Chap. 2 established, showed the inability of governance to finance urban renewal needs already apparent a decade or two before (Chap. 1). The instance of Bushwick's needs was apparent when sections of it were declared 'preventive renewal areas' 5 years before the arson fires of 1977. The witnessing of the incident was an awakening to governments to the absence of policies that may have prevented the demise of Bushwick.

7.3 Bushwick and Northeast Brooklyn Planning Directions

Stages of Bushwick's urban history, from hay days (Chap. 1) to demise (Chap. 2) to recovery (Chap. 5) showed the slippery curve of keeping communities well serviced. This servicing relied on a strong foundation of housing security, community safety and accessible jobs. Bushwick's early development (from formation of Bushwick Town in mid-1800s) (Chap. 1) was in stark contrast to the decline in the 1960s (Chap. 2). It was the dramatic demise (1977 arson fires awakening) that opened up a vast canyon of urban planning challenges that came in the aftermath. The recovery, with deep commitments from all sectors and continuing strengthening of leadership, became crucial to tackling the needs (Chap. 5).

The event of 1977 resulted in a call for community and housing development (Chap. 2). The pain staking attention to that recovery started to show results in the 1980s, 1990s and early 2000s. One key to that recovery was the successful application of the Brownfields Redevelopment Plan instrument to the West Bushwick Rhinegold site (Chap. 5). Likewise, looking beyond Bushwick, the application of the 197-a Plan process showed the strength of that planning instrument as applied to Williamsburg waterfront, Greenpoint and the combined Greenpoint/Williamsburg district. The strength of this instrument appeared to be the community engagement basis of the planning process. There were questions raised however (Chap. 4) about the City of New York staying with the program to complete all the tasks as identified in the plans (including mustering agency cooperation). It can be concluded that within Bushwick the application of the brownfield restoration planning program produced considerable success. Likewise the general successes (but with further attention needed) of 197-a Plan processes in Northeast Brooklyn neighborhoods of Williamsburg and Greenpoint further anchored the recovery of other close-by neighborhoods such as Bushwick.

The success of planning instruments (Chaps. 4 and 5) was vital to assisting the revitalization of neighborhoods of Bushwick and Northeast Brooklyn. Examining two other prime Brooklyn neighborhoods showed the contrast in planning instruments applied by the City. Comprehensive redevelopment planning (including master plan making) is the thrust of planning at Brooklyn Downtown and Southeast Brooklyn (including Coney Island). Given the public infrastructure in these two neighborhoods, the commitment of the City and the business sectors in these two areas is evident. The City of New York's planning role here is as a lead player, with the financing of public developments (in particular education and arts) stimulating private investments, aims at overall urban renewal of these neighborhoods.

It is the business sector (via the work in particular of the Downtown Brooklyn Partnership) that continues to drive the significant planning and investment cycles behind the recovery of Downtown Brooklyn (Chap. 5). The strengthening of that district has a flow on to neighborhoods such as Bushwick, Greenpoint and Williamsburg. Those neighborhoods are all within easy public transport access to Brooklyn Downtown and maintain strong historical, economic and cultural ties with Downtown. Finally, the demise of Coney Island was a shock for all New Yorkers over several decades. Hence the planning incentives as outlined (Chap. 5) appear to be widely embraced by all who know Coney Island for the attraction it held for over a century. The question posed in Chap. 3 was how do communities and governments move to sustainable urban planning? This subject is focused on next.

7.4 Sustainable Urban Planning (SUP) Directions

Governments in most places across the globe are examining more closely principles of ecologically sustainable development (ESD), as outlined in Chap. 3. Sustainable urban planning (SUP) is about respecting the three parts of sustainability (environment, social/cultural and economic). Most people today accept that the 'environment' part of sustainability needs to be the core of the circle, thus most critical of the three components (different models discussed in Chap. 3). The chapter showed the advantages of governments around the world adopting policies of conservation of resources (especially in urban areas). The challenge here is the means of incorporating sustainability principles within urban planning practices. In looking at the sustainable urban planning (SUP) directions initiated by the City of New York and New York State (state interest especially within the environment protection and alternative energy sectors), both levels of governance appeared to be moving in that direction, though a distance yet to go.

The City of New York community boards, it is concluded, offer a foundation for sustainable urban planning. With the sharing experiences among community boards in the planning steps of creating a 197-a Plans (Chap. 4). For example, there is considerable potential for boards to adopt sustainability criteria (with agreed measurement standards) within their operations. As illustrated in Chap. 3, such an approach by a community could result in 'sustainability score cards' and 'sustainability report cards' being applied. Taking these steps would likely lead the City and boards to adopt the required programs that reflected a measured way to improve communities. Reflecting back to Bushwick's (or Coney Island's) demise, one imagines that were sustainability reporting a part of the administrations at the time (1960s and 1970s) the extent of the decline could have been abated, or even averted.

The strength of a sustainable urban planning (SAP) approach is also reflected in the universality of that approach. As illustrated in Chap. 3, it is the United Nations and international non-government organizations (NGOs) that have set agendas, education programs and measured standards to assist communities to become more sustainable. The work of the International Center for Local Environment Initiatives (ICLEI), was noted in Chap. 5, (and especially working with local government bodies). This work could be one model of expanding a government's efforts to assist districts, neighborhoods, or a whole city to work towards better and more secure living environments.

Planning instruments such as City of New York brownfield redevelopment planning and the 197-a Plan process could adopt a range of sustainability indicators to assist the decision making in setting priorities for subsequent programs and budgets. Crucial to sustainable urban planning is the engagement of the community and sustaining that engagement for the duration of any programs. The urban planners quoted in Chap. 4 made a common observation of plans being adopted by the City of New York and then only minimum actions being taken to implement those plans by the City or its agencies (as noted above). Overall, it can be concluded that a degree of progress has been made in the City of New York taking steps to meet criteria of sustainability within urban planning. As the City has illustrated in its planning initiatives in Brooklyn to date (Chaps. 4 and 5), the foundation to create sustainable communities is already partly in place. It is critical that the City (and State) monitor the progress of areas that are recovering (i.e. Bushwick and Coney Island) from decline.

Cast back to the first formation of Old Brooklyn's Towns (including Bushwick Town and areas Greenpoint and Williamsburg) in 1854 (Chap. 1). The zeal to build prosperous and socially inclusive neighborhoods in New York (over century and a half) should continue for centuries to come. The foundations for experimental education aimed to better prepare for the challenges of urban futures (as canvassed in Bushwick high schools in Chap. 6) needs now to be looked at for wider applications. The universalists writers (Whitman, Thoreau, Blavatsky, George, H Miller and A Miller) associated with New York City as introduced in the book's 'dedication' could well have been prescient about the need for sustainability approaches in building livable communities (thus respecting nature). Recent calls by Al Gore and others (Chap. 3) (and President Obama addressing climate change recently) for continued work in adopting a sustainable urban planning (SUP) approach would have been welcome by all the above writers. We have in 2013 a historical reminder of supreme leadership, with the 150th anniversary of President Lincoln's Gettysburg Address (11 Nov 1863), and the 50th anniversary of the assassination (22 Nov 1963) of the popular American President, John F. Kennedy. Both continue to inspire generations.

We can all thank New York City for giving us the opportunity to prepare for urban futures, especially through the experiences and lessons learned in Bushwick, Greenpoint, Williamsburg, Downtown Brooklyn, and Southeast Brooklyn (including Coney Island) as examined in these chapters. One main lesson seems to be repeated historically, retain the best of the past and conserve the most for the future generations.

Appendices

Appendix 1: Class of 1961 Bushwick High School Participants in 'Bushwick Story' (Plate A1)

Appendix 2: Bushwick Family Sketch: Rauscher's 1900–2010

Each Bushwick family over five (4) generations (say 1900–2010) would tell an intriguing story about growing up in Bushwick. Here is a profile of family life in Bushwick, the Rauscher family. The years 1900–1949 are covered in Plate A2 and the years 1953–2010 are covered in Plate A3. A brief summary follows.

Plate A2 (Left to Right photos)

1900

In 1900 the Rauscher's joined a business venture in boat building with the Scaife's. Here Rauscher & Scaife plied their trade at Rockaway, Brooklyn, into the 1900s. Boat building was a major industry in Brooklyn where good harbors afforded protection and timber for boats was readily available from Up State NY.

1915

In 1915 timber was still the main building material for boats as well as housing. Here Carl Rauscher (grandfather of Ray Rauscher (author) hoists his son, Henry Rauscher (author's father). The house was in Staten Island, an early Rauscher residence. Note the chicken pen and vegetable garden, a common feature in most backyards at that time.

R.C. Rauscher and S. Momtaz, *Brooklyn's Bushwick - Urban Renewal in New York, USA: Community, Planning and Sustainable Environments*, DOI 10.1007/978-3-319-05762-0,

Plate A1 Class of Bushwick HS 1961 and Principal of HS 1961. *Left* to *Right*: Mari Degender (nee Grossman), Susan Ferrandiz, Betty Lee (nee Wasielewski), Tina Morello, Elaine Portnoy (nee Grossman), Ray Rauscher, Pat Ward, Dr. Clara A Molendyk, Principal. Bushwick HS 50s/60s (Source: Oriole 1961 Graduation Book, Bushwick High School)

Plate A2 Bushwick Family Sketch: Rauscher's 1900–1949. *Left* to *Right*: 1900 Rockaway, Rauscher's Boat Building; 1915 Station Island Carl Rauscher and Son Henry, back yard; 1925 Menahan St, Henry Rauscher, back yard; 1930 Roselands Dance Hall (Manhattan), Henry Rauscher, Small Band Leader; 1945 (WW2) Arthur Rauscher (left) and Henry Rauscher, brother; 1949 Rauscher's (l–r) from back Sonny, Grace (mom), Henry (dad), Ray, Arthur, Gracy and Joyce (Source: Ray Rauscher)

Plate A3 Bushwick Family Sketch: Rauscher's 1953–2010. *Left* to *Right*: 1953 Amelia Rauscher, Menahan St Home; 1954 Cedar St (home) Bushwick 1954; 1955 Broadway, Bushwick (Ray); St Barbara's, Family Catholic Church, Central Ave; 1961 Bushwick High School, Family School; Class of 1961; 2010 Ray Rauscher (Source: Ray Rauscher)

1925

By 1925 the Rauscher family had moved to Bushwick. Backyards in the neighborhood were sparsely laid out. Here Henry Rauscher (10 years old) plays on a timber horse in the midst of a swing set and shed in Menahan St, Bushwick.

1930

In 1930 the big bands were the draw card all over the USA. Here Henry Rauscher is photographed in full band tux with his drums. Henry, like so many Brooklyn music industry people in that era of bands, started up a five (5) piece band at Roselands, the Manhattan mecca for dance and music (located in mid-Manhattan and still in 2010 was functioning as a music venue).

1945

The Second World War affected all Brooklyn families, with the Brooklyn Navy Yard producing a major component of the US frigate demands. Here Henry Rauscher joins his brother Arthur Rauscher in Menahan St, Bushwick (home of their parents). Henry's son, Sonny Rauscher, would later serve in the Korean War.

1949

Families in the 1940s were larger than today, a flashback to the Rauscher family in 1949.

Plate A3 covers the time span 1953–2010 (left to right photos).

1953

By 1953 Menahan Street home of the Henry's parents, Carl and Amelia, would become the family gathering place for all occasions. This house (Amelia in front with Storm, German Sheppard dog) is typical wrought iron gate, cellar level and two stories above, with entries below as well as steps to top floor. Note the decorations to celebrate Armistice Day, one of many celebrations that brought Brooklyn families at the time together.

1954

In 1954 the Henry and Grace Rauscher home was in Cedar Street, Bushwick, just off Bushwick Av and a dozen streets from Henry's parents Carl and Amelia's Menahan home. Henry and Grace had a total of seven (7) children, staggered between an older group of Lorraine, Sonny and Billy and a younger group of Gracie, Arthur, Ray and Joyce. Brooklyn families often lived close to their relatives for extended family support and social get-togethers. Here in Cedar St, Ray Rauscher carried a paper bag back from the corner shop, brother Arthur just visible behind Ray. Note the cars (Dodge, Plymouths, Desoto, Fords, etc.) in the street. These cars would sit there all week at times and be used on weekends for shopping or recreation (such as a trip to Coney Island or Jones Beach, Long Island).

1955

In 1955 Boy Scouts and other community organizations were still popular with families in Brooklyn. Here Ray (Troop 13 Brooklyn) stands beside a corner shop off Broadway, Bushwick. Note all the cigarette ads at the time on the tobacco shop.

1957

St Barbara's Church was the center of a good deal of community life for Bushwick families.

1961

Bushwick High School was the high school that the Rauscher children attended. Here was the school when the class of 1961 graduated.

2010

By 2010 a good deal had changed in Brooklyn, Bushwick and the world. Here Ray stands beside a fig tree in inner city of Sydney, where he currently lives. The world too had changed to full photo digital color. Ray contributes this poem to all families keeping memories of early times.

Appendix 3: Selected Urban Planning Websites

Web links last referenced: 20 Nov 2013

The following web sites are considered good sources of information and contacts for urban planning and sustainability stemming from work of groups and institutions mostly located in Brooklyn (including Bushwick) and New York City.

93 Nevins Street
www.93nevins.com/
The web is about the development of a green house in Boerum Hill, Brooklyn.

Academy of Urban Planning
http://www.aupnyc.org
Home page for Academy of Urban Planning located in Bushwick (one of three schools replacing Bushwick HS).

Arts in Bushwick
www.artsinbushwick.org
Group that promotes the arts, including running an annual Bushwick arts festival

Bedford Stuyvesant
http://www.nyc.gov/html/hpd/html/about/bed-stuy-walking-tour.shtml
Walking tour of urban improvements

Billburg
www.billburg.com
To help people keep up to date on events in Williamsburg.

Bio-thinking
www.biothinking.com/greendrinks
International in scope, this is a social networking forum for those who work in the environmental field.

Brooklyn: About.com
www.brooklyn.about.com
Blog potpourri about Brooklyn.

Brooklyn Center for the Urban Environments (BCUE)
As of April 09 the BCUE was not in operation. The latest news about the Center can be found at Sustainable Flatbush www.sustainableflatbush.org

Brooklyn Historical Society
http://www.brooklynhistory.org/visitor/visitors.html
Located in the Brooklyn Heights historic district, exhibits cover all aspects of Brooklyn's rich history.

Brooklyn on Line
www.brooklynonline.com
Web is a large collection of Neighborhood resources including photos, history, events, etc.

Brooklyn Paper
www.brooklynpaper.com
This offers an online paper which covers news of western Brooklyn from Greenpoint to Gravesend.

Community Energy: Wind Energy Leaders
www.newwindenergy.com/
Commercial company involved with providing wind energy to businesses and individuals.

Community Environmental Center
www.cecenter.org/
A non-profit organization dealing with energy reduction and green building solutions.

Green Buildings NYC
www.greenbuildingsnyc.com
Purpose is to connect green buildings, business and real estate to the law

Harlem Renewal – walking tour.
http://www.nyc.gov/html/dcp/html/mxb/index.shtml
The search button will take you to the walking tour

International center for Local Environment Initiatives (ICLEI)
www.iclei.org
Promotes initiatives in environmental improvements at the local level

Make the Road Count
www.maketheroad.org
Founded in 1997 in Bushwick, the organization promotes equal rights and economic and political opportunity for low income New Yorkers through community organizing.

Metrotech BID Downtown Brooklyn
www.metrotechbid.org
Metrotech is non-profit community development organization founded in 1992 to further revitalization of Downtown Brooklyn.

Municipal Arts Society
www.mas.org
Contains the document, "Planning for all New Yorkers: an atlas of community based plans-Brooklyn."

Newtown Creek Alliance
www.newtowncreekalliance.org
The alliance is a community organization dedicated to revitalizing, restoring and revealing Newtown Creek which is on the border of Bushwick and Greenpoint

Neighborhood Preservation Center
www.neighborhoodpreservationcenter.org

New Village
http://www.newvillage.net

New York City (City Council) www.home.nyc.gov

New York City Department of City Planning
www.home.nyc.gov/html/dcp/home.html
New York City site with major planning projects (search button for example 'Coney Island')

New York City Department of Housing Preservation and Development – Bushwick
www.nyc.gov/html/hpd/html/about/bushwick-walking-tour.shtml
Contains a walking tour of new urban developments in Bushwick

New York State Environmental Research and Development Authority (NYSERDA)
www.nyserda.org
Promotes innovations in environmental advances (i.e. energy conservation)

Planners Network: the Organization of Progressive Planning
www.plannersnetwork.org
This is an organization for those interested in urban planning.

Pratt Center for Community Development
www.picced.org
The Pratt Center is located within the Pratt Institute.

Project for Public Spaces
www.pps.org/
PPS is a non-profit organization dedicated to helping people create and sustain public places that build communities.

Ridgewood-Bushwick Senior Citizens Corp
www.rbscc.org
RBSCC is a not for profit group that provides a range of programs, including housing developments, in the Bushwick and Ridgewood neighborhoods

Stable Brooklyn Community Group
www.stablebrooklyn.com
Contributing to sensible Neighborhood design

Up in Flames
www.brooklynhistoricalsociety.org

Wikipedia on Bushwick
http://en.wikipedia.org/wiki/Bushwick
Site gives a significant background to Bushwick with numerous links.

Post Script

As the story opened with a dedication to Walt Whitman, the story closes with a poem to Whitman (Plate PS1).

With the Obama presidential second term in place, the administration in Washington is facing a barrage of urban planning issues. A mural depicting a call for perfection in the American union is duplicated in Plate PS2, a perfection needed in moving to sustainable communities across America and around the world. The mural is at (2010) Grattan, between Morgan and Bogart St, Brooklyn.

> We the people, in order to form a more perfect union....two hundred and twenty one years ago…a group of men gathered and, with these simple words, launched America's improbably experiment in democracy…The document they produced was eventually signed but ultimately unfinished. It was stained by this nation's original sin of slavery…I believe deeply that we cannot solve the challenges of our time unless we solve them together… unless we perfect our union by understanding that we may have different stories, but we hold common hopes… And if we walk away now, if we simply retreat into our respective corners, we will never be able to come together and dissolve challenges like health care, or education, or need to find good jobs for every American. But it is where we start. It is where our union grows stronger. And as so many generations have come to realize, over the course of the two-hundred and twenty one years since a band of patriots signed that document in Philadelphia. That is where the perfection begins
>
> Barrack Obama

You can read the whole speech here: www.my.barackobama.com/page/content/hisownwords

Finally, a closing note is included as a thank you to the students of the Academy of Urban Planning (Bushwick). May the school and its pupils continue to meet success. A success in the schooling system can compliment successes in building sustainable communities. The pioneers of Brooklyn and Bushwick, as reviewed in the first part of the book, would be proud as Bushwick strides to renewal. A final tribute to all of Brooklyn's Bushwick residents, past and present (Plate PS3).

R.C. Rauscher and S. Momtaz, *Brooklyn's Bushwick - Urban Renewal in New York, USA: Community, Planning and Sustainable Environments*, DOI 10.1007/978-3-319-05762-0,

You have given us great insight into the depths of life
Your Leaves of Grass spring of eternal light
You have lived so many lives in your term on earth
You embrace all men, women, child and leaf

Your enthusiasms for living knows no equal on earth
You have covered every blade of grass pushing through
You have nudged us to sort the complexities of life
You view life as continual jumping of logs down a raging river

Your passion still infuses every soul of salt
Music and harsh reality rest side by side in balance
You spell out all surprises and disappointments
Always the full life and unexpected trials coming

Mr Whitman, your work is not weathered with age or taste
You're true for today's masses alive as of yesterday
More so, your potion would sooth lives so empty today
Woes of many would be comforted with your humanity

The rock you stood on gave strength to people then
Hopes uplifted and burning desire were ignited
The road of life you pointed to flows with ease and pulse
Words inscribed for thirsty today and fifty centuries to come

We lost you Mr Whitman on our road so busy surviving
Rushing and repairing, maintaining and investing
Somewhere the rhythm of life gets mixed and disparate
The total results dissipated in lost directions today

In your Blades of Grass you give a compendium of life
How much we grab and possess is within our grasp
There is no boundary to that exuberance you offer
We will not lose sight of your wisdom love of city living

Plate PS1 Whitman of Brooklyn (Source: Ray C. Rauscher (Unpublished))

Plate PS2 The perfection begins (Source: The www.BushwickBK.com Photo (#371))

The Bushwick urban renewal story takes us to mind maps etched deep
We text the passages of fading streets of bustle and heartfelt pangs
We tempt travel again in flickering sparks alighting under trolley wires
Haze lifts pushing curtains heavy into peaks of past still buried

For within ourselves permission granted to fly into map corners precious
Laid out still are footprints of snow never melted in sneakers deep
Blanket again that white fine fluff swept to corners of your soul
Relive those secrets and friendships bonded in Bushwick etchings

Plate PS3 Poem: Maps Etched Deep in Bushwick (Source: R Rauscher (unpublished))

Author Biographies

Raymond Charles Rauscher

http://www.ned.org.au/sdn/sdnstories/keepinginspiration.htm

I was born (1943) in Menahan St, Bushwick, and lived early years in Cedar St (1945–1958). Bushwick. High School years were spent in the adjoining neighborhood of Ridgewood, Queens (1958–1961), attending Bushwick High School. University years were spent in the Bronx (1961–1965), attending City College of New York (1966 BE – Civil) (uptown Manhattan). A tour of duty was completed in Vietnam (1966–67), engaged in engineering projects (Corps of Engineers, Highlands Vietnam).

In 1968 the first part of a masters urban planning degree was completed at the University of Michigan. The research study here was *Urban Crisis: Proposal for Region Serving New Cities* (Rauscher 1969) (unpublished), a case study of Livingston County (outside Detroit). Subsequent 11 years in Australia (married to Diane, two children Anna and Maree) led to a Masters Town and Country Planning (thesis *Community Response to a Redevelopment Proposal* ((Sydney University (1971)). In 2009 I was awarded a doctoral from the University of Newcastle, Australia (thesis

R.C. Rauscher and S. Momtaz, *Brooklyn's Bushwick - Urban Renewal in New York, USA: Community, Planning and Sustainable Environments*, DOI 10.1007/978-3-319-05762-0,

Sustainable Area Planning Framework for Ecologically Sustainable Development, Case Study Wyong Shire, NSW, Australia). http://ogma.newcastle.edu.au:8080/vital/access/manager/Repository/uon:4306.

My initial reaction on touring Bushwick (visiting parents and sibling families) in 1979 (2 years after the Bushwick arson fires) convinced me that urban renewal here faced an enormous challenge. Subsequent visits in 1982, 1993, 1998, 2000, 2007, 2008, and 2010 enabled further research to be completed, focusing on urban renewal in Bushwick and Northeast Brooklyn (including Greenpoint and Williamsburg), and with visits to Brooklyn Downtown and Southeast Brooklyn (including Coney Island).

Dr. Rauscher is currently a Conjoint Lecturer at the University of Newcastle, Australia. He is also a director of Habitat Association for Arts and Environment Inc. (www.habitatassociation.com.au). Ray has lived in and been active in community development and planning in localities, such as: USA, including Brooklyn and Denver; Australia, including Sydney (Ashfield, Canterbury, Hornsby, North Sydney, St George and Marrickville), Central Coast Region (Gosford City and Wyong Shire); and, Hunter Region (Newcastle City and Singleton).

Salim Momtaz

I was born in Dhaka, the capital city of Bangladesh. I received my B.Sc. with Honors and Masters in Geography from the University of Dhaka. Urban Geography and Urban Planning were two courses in my Masters curriculum that introduced me to the concepts of urban planning models and urban renewal.

Between 1986 and 1990 I was at the University College London under a Commonwealth Scholarship doing PhD in Regional Planning. I migrated to Australia in 1994 and started teaching at Central Queensland University and later joined the University of Newcastle. I taught a course titled 'Planning for Sustainability' which gave me the opportunity to learn more about sustainable urban planning models and disseminate my knowledge to undergraduate students. Ray and I become interested about publishing a book on Bushwick and Northeast

Brooklyn urban renewal when we were working on our first book (see later). I was amazed by the amount of information Ray had collected on urban history and transformation of Bushwick during his numerous visits to New York between 1979 and 2010. His family history and its link with the history of Bushwick and Northeast Brooklyn made this book project a possibility. I felt there was a good deal to share with people of that district and beyond. I thus happily shared my knowledge and skills to bring the idea of the book to fruition.

Dr Salim Momtaz *is a Senior Lecturer at the University of Newcastle, Australia. He teaches in the area of Sustainable Resource Management. Salim's current research interests include: development and environment; climate change adaptation; environmental governance; and, social impact assessment. He had a stint in the US teaching Environmental and Social Impact Assessment at Georgetown University, Washington D.C., as a Visiting Professor. He has also been a Rotary International Ambassadorial Fellow to Bangladesh. Finally, Salim has been a member of the Scientific Advisory Committee, Netherlands Government Research Organization, between 2007 and 2010. His two recent books are: Evaluating Environmental and Social Impact Assessment (Elsevier 2013) (co-author S. Kabir), and Sustainable Communities: A Framework for Planning (Springer 2013) (co-author with R. Rauscher)* http://www.springer.com/environment/sustainable+development/book/978-94-007-7508-4.

Glossary

Act An act is "an order, law or judgment as of Parliament" (Macquarie Dictionary 2005).

Agenda 21 is a detailed plan of actions dealing with all aspects of ecologically sustainable development and desirable national policies. The concept was agreed to by the national representatives at the United Nations Conference on Environment and Development (UNCED) at Rio de Janeiro in June 1992.

Agreement coming to an arrangement; a condition of agreeing. (Macquarie Dictionary 2005)

Area A region (or other size) such as a settled *area*; an area of land delineated by geographical features; and, can also be an administrative area such as a sub-part of a city (Macquarie Dictionary 2005).

Climate Change global changes in climate associated with the greenhouse effect, including the overall effects on climate of human made and natural changes;

Cultural Land relating to tradition of indigenous people lands (Macquarie Dictionary 2005)

Ecology branch of biology dealing with organism and their environment

Environment the aggregate of surrounding things, conditions or influences. Broad natural surrounding conditions, such as the bush, rivers, air, sea in which human and natural elements exist (Macquarie Dictionary 2005).

ESD (Ecologically Sustainable Development) using, conserving and enhancing the community's natural resources so that ecological processes on which life depends are maintained and the total quality of life, now and into the future, can be increased.

Indicator recording variations, reactions or changes affecting a system.

Kyoto Agreement under the United Nations as signed by nation; effective when Russia signed in September 2004; requires nations to reduce greenhouse gases

Limits to Growth the title of a book published in 1972 by the Club of Rome with predictions of severe consequences if the world's population and resource use continued to grow

R.C. Rauscher and S. Momtaz, *Brooklyn's Bushwick - Urban Renewal in New York, USA: Community, Planning and Sustainable Environments*, DOI 10.1007/978-3-319-05762-0,

Local Environment an environment limited to a particular place or small area; relates to parts of an area as of a system.

Natural Capital potential wealth in resources and the environment due to their original natural qualities, as against human made capital due to human activities

Neighborhood a district or locality with reference to living boundaries.

Precinct a space of definite or understood limits and its environs (Macquarie Dictionary 2005).

Principle a principle is 'a rule or law exemplified in natural phenomena' (The Macquarie Dictionary 2005).

Protocol a protocol is 'rules of behaviour to be agreed upon by heads of organizations' (Macquarie Dictionary 2005).

Seniors Area a planning term relating to areas set aside for older aged group living

Sensitive affected by external agencies or influences; affected by small amounts of change (Macquarie Dictionary 2005).

Standard a basis of comparing things of a similar nature; serving as a basis or measure or value (Macquarie Dictionary 2005).

Suburb a part of a district or local government area

Sustainable Urban Planning (SUP) urban planning that includes the adoption of ecological sustainable development (ESD) principles

Sustainability managing our natural resources in a way that maintains their environment, economic and cultural values, so that they continue to be available in the long-term

Urban relation to a city or town (Macquarie Dictionary 2005)

Vulnerable susceptible to physical impact; not protected or immune

Index

R.C. Rauscher and S. Momtaz, *Brooklyn's Bushwick - Urban Renewal in New York, USA: Community, Planning and Sustainable Environments*, DOI 10.1007/978-3-319-05762-0,

MIX
Papier aus verantwortungsvollen Quellen
Paper from responsible sources
FSC® C105338

If you have any concerns about our products, you can contact us on
ProductSafety@springernature.com

In case Publisher is established outside the EU, the EU authorized representative is:
Springer Nature Customer Service Center GmbH
Europaplatz 3, 69115 Heidelberg, Germany

Printed by Libri Plureos GmbH
in Hamburg, Germany